SUCCESSFUL RESPONSE STARTS WITH A MAP

Improving Geospatial Support for Disaster Management

Committee on Planning for Catastrophe:
A Blueprint for Improving Geospatial Data, Tools, and Infrastructure

Mapping Science Committee

Board on Earth Sciences and Resources

Division on Earth and Life Studies

NATIONAL RESEARCH COUNCIL
OF THE NATIONAL ACADEMIES

**Library
Quest University Canada
3200 University Boulevard
Squamish, BC V8B 0N8**

THE NATIONAL ACADEMIES PRESS
Washington, D.C.
www.nap.edu

THE NATIONAL ACADEMIES PRESS 500 Fifth Street, N.W. Washington, DC 20001

NOTICE: The project that is the subject of this report was approved by the Governing Board of the National Research Council, whose members are drawn from the councils of the National Academy of Sciences, the National Academy of Engineering, and the Institute of Medicine. The members of the committee responsible for the report were chosen for their special competences and with regard for appropriate balance.

This study was supported by the National Aeronautics and Space Administration, Award No. W-92759; U.S. Department of Commerce/National Oceanic and Atmospheric Administration, Contract No. 50-DGNA-1-90024; U.S. Department of Defense/National Geospatial-Intelligence Agency, Award No. NMA501-03-1-2019 T0029; and Department of the Interior/U.S. Geological Survey, Grant No. 03HQGR0147. Any opinions, findings, conclusions, or recommendations expressed in this publication are those of the authors and do not necessarily reflect the views of the organizations or agencies that provided support for the project.

International Standard Book Number-10: 0-309-10340-1
International Standard Book Number-13: 978-0-309-10340-4

Additional copies of this report are available from the National Academies Press, 500 Fifth Street, N.W., Lockbox 285, Washington, DC 20055; (800) 624-6242 or (202) 334-3313 (in the Washington metropolitan area); Internet, http://www.nap.edu

Cover: Designed by Michele de la Menardiere. Top left shows a U.S. Coast Guard rescue from a home surrounded by floodwaters after Hurricane Katrina in New Orleans (AP photo by David J. Phillip); middle right, a simulation of a category-3 storm surge in New Orleans showing emergency services (image courtesy the National Geospatial-Intelligence Agency); bottom left shows an area in Long Beach, Mississippi, roughly 1 week after Hurricane Katrina (image, courtesy of Bruce Davis, Department of Homeland Security). Background shows satellite imagery of a forest fire (image courtesy Digital Globe).

Copyright 2007 by the National Academy of Sciences. All rights reserved.

Printed in the United States of America

THE NATIONAL ACADEMIES
Advisers to the Nation on Science, Engineering, and Medicine

The **National Academy of Sciences** is a private, nonprofit, self-perpetuating society of distinguished scholars engaged in scientific and engineering research, dedicated to the furtherance of science and technology and to their use for the general welfare. Upon the authority of the charter granted to it by the Congress in 1863, the Academy has a mandate that requires it to advise the federal government on scientific and technical matters. Dr. Ralph J. Cicerone is president of the National Academy of Sciences.

The **National Academy of Engineering** was established in 1964, under the charter of the National Academy of Sciences, as a parallel organization of outstanding engineers. It is autonomous in its administration and in the selection of its members, sharing with the National Academy of Sciences the responsibility for advising the federal government. The National Academy of Engineering also sponsors engineering programs aimed at meeting national needs, encourages education and research, and recognizes the superior achievements of engineers. Dr. Wm. A. Wulf is president of the National Academy of Engineering.

The **Institute of Medicine** was established in 1970 by the National Academy of Sciences to secure the services of eminent members of appropriate professions in the examination of policy matters pertaining to the health of the public. The Institute acts under the responsibility given to the National Academy of Sciences by its congressional charter to be an adviser to the federal government and, upon its own initiative, to identify issues of medical care, research, and education. Dr. Harvey V. Fineberg is president of the Institute of Medicine.

The **National Research Council** was organized by the National Academy of Sciences in 1916 to associate the broad community of science and technology with the Academy's purposes of furthering knowledge and advising the federal government. Functioning in accordance with general policies determined by the Academy, the Council has become the principal operating agency of both the National Academy of Sciences and the National Academy of Engineering in providing services to the government, the public, and the scientific and engineering communities. The Council is administered jointly by both Academies and the Institute of Medicine. Dr. Ralph J. Cicerone and Dr. Wm. A. Wulf are chair and vice chair, respectively, of the National Research Council.

www.national-academies.org

COMMITTEE ON PLANNING FOR CATASTROPHE: A BLUEPRINT FOR IMPROVING GEOSPATIAL DATA, TOOLS, AND INFRASTRUCTURE

MICHAEL F. GOODCHILD, *Chair*, University of California, Santa Barbara
ANDREW J. BRUZEWICZ, U.S. Army Corps of Engineers, Remote Sensing/GIS Center, Hanover, New Hampshire
SUSAN L. CUTTER, University of South Carolina, Columbia
PAUL J. DENSHAM, University College London
AMY K. DONAHUE, University of Connecticut, West Hartford
J. PETER GOMEZ, Xcel Energy, Denver, Colorado
PATRICIA HU, Oak Ridge National Laboratory, Knoxville, Tennessee
JUDITH KLAVANS, University of Maryland, College Park
JOHN J. MOELLER, Northrop Grumman TASC, Chantilly, Virginia
MARK MONMONIER, Syracuse University, New York
BRUCE OSWALD, James W. Sewell Co., Latham, New York
CARL REED, Open Geospatial Consortium, Inc., Ft. Collins, Colorado
ELLIS M. STANLEY, SR., Emergency Preparedness Department City of Los Angeles, California

Staff

ANN G. FRAZIER, Program Officer
JARED P. ENO, Senior Program Assistant (since August 2006)
AMANDA M. ROBERTS, Senior Program Assistant (through August 2006)

MAPPING SCIENCE COMMITTEE

KEITH C. CLARKE, *Chair*, University of California, Santa Barbara
ISABEL F. CRUZ, University of Illinois, Chicago
ROBERT P. DENARO, NAVTEQ Corporation, Chicago, Illinois
SHOREH ELHAMI, Delaware County Auditor's Office, Delaware, Ohio
DAVID R. FLETCHER, GPC, Inc., Albuquerque, New Mexico
JIM GERINGER, ESRI, Wheatland, Wyoming
JOHN R. JENSEN, University of South Carolina, Columbia
NINA S.-N. LAM, Louisiana State University, Baton Rouge
MARY L. LARSGAARD, University of California, Santa Barbara
DAVID R. MAIDMENT, The University of Texas, Austin
ROBERT B. MCMASTER, University of Minnesota, Minneapolis
SHASHI SHEKHAR, University of Minnesota, Minneapolis
NANCY TOSTA, Ross & Associates Environmental Consulting, Ltd., Seattle, Washington
EUGENE TROBIA, Arizona State Land Department, Phoenix

Staff

ANN G. FRAZIER, Program Officer
JARED P. ENO, Senior Program Assistant (since August 2006)
AMANDA M. ROBERTS, Senior Program Assistant (through August 2006)

BOARD ON EARTH SCIENCES AND RESOURCES

Members

GEORGE M. HORNBERGER, *Chair*, University of Virginia, Charlottesville
M. LEE ALLISON, Arizona Geological Survey, Tucson
GREGORY B. BAECHER, University of Maryland, College Park
STEVEN R. BOHLEN, Joint Oceanographic Institutions, Washington, D.C.
KEITH C. CLARKE, University of California, Santa Barbara
DAVID COWEN, University of South Carolina, Columbia
ROGER M. DOWNS, Pennsylvania State University, University Park
JEFF DOZIER, University of California, Santa Barbara
KATHERINE H. FREEMAN, Pennsylvania State University, University Park
RHEA L. GRAHAM, Pueblo of Sandia, Bernalillo, New Mexico
ROBYN HANNIGAN, Arkansas State University, State University
MURRAY W. HITZMAN, Colorado School of Mines, Golden
V. RAMA MURTHY, University of Minnesota, Minneapolis
RAYMOND A. PRICE, Queen's University, Ontario, Canada
BARBARA A. ROMANOWICZ, University of California, Berkeley
JOAQUIN RUIZ, University of Arizona, Tucson
MARK SCHAEFER, Global Environment and Technology Foundation, Arlington, Virginia
RUSSELL STANDS-OVER-BULL, BP American Production Company, Pryor, Montana
BILLIE L. TURNER II, Clark University, Worcester, Massachusetts
TERRY C. WALLACE, JR., Los Alamos National Laboratory, New Mexico
STEPHEN G. WELLS, Desert Research Institute, Reno, Nevada
THOMAS J. WILBANKS, Oak Ridge National Laboratory, Tennessee

Staff

ANTHONY R. DE SOUZA, Director
PAUL M. CUTLER, Senior Program Officer
ELIZABETH A. EIDE, Senior Program Officer
DAVID A. FEARY, Senior Program Officer
ANNE M. LINN, Senior Program Officer
ANN G. FRAZIER, Program Officer

SAMMANTHA L. MAGSINO, Program Officer
RONALD F. ABLER, Senior Scholar
VERNA J. BOWEN, Administrative and Financial Associate
JENNIFER T. ESTEP, Financial Associate
CAETLIN M. OFIESH, Research Associate
JARED P. ENO, Senior Program Assistant
NICHOLAS D. ROGERS, Senior Program Assistant

Acknowledgments

This report has been reviewed in draft form by individuals chosen for their diverse perspectives and technical expertise, in accordance with procedures approved by the National Research Council's (NRC's) Report Review Committee. The purpose of this independent review is to provide candid and critical comments that will assist the institution in making its published report as sound as possible and to ensure that the report meets institutional standards for objectivity, evidence, and responsiveness to the study charge. The review comments and draft manuscript remain confidential to protect the integrity of the deliberative process. We wish to thank the following individuals for their review of this report:

Massoud Amin, University of Minnesota
Jane Bullock, Bullock and Haddow, LLC
Michael Domaratz, U.S. Geological Survey (retired)
Gerald Galloway, University of Maryland
David Kehrlein, Environmental Systems Research Institute
Arthur Lerner-Lam, Columbia University
Henk Scholten, Free University in Amsterdam
Seth Stein, Northwestern University
Gayle Sugiyama, Lawrence Livermore National Laboratory

Although the reviewers listed above have provided many constructive comments and suggestions, they were not asked to endorse the conclusions or recommendations, nor did they see the final draft of the report before its release. The review of this report was overseen by Dr. Robert

Hamilton, National Research Council (retired) and U.S. Geological Survey (retired), and Dr. Chris G. Whipple, ENVIRON International Corporation. Appointed by the NRC, they were responsible for making certain that an independent examination of the report was carried out in accordance with institutional procedures and that all review comments were carefully considered. Responsibility for the final content of this report rests entirely with the authoring committee and the institution.

Preface

After the events of September 11, 2001, there was a widespread sense in the United States and in many other parts of the world that humanity was entering a new and more dangerous era. Subsequent events, such as the Indian Ocean tsunami of 2004, the Gulf Coast hurricanes of 2005, and the terrorist bombings of July 7, 2005, in London have if anything strengthened that feeling, as have the potential threats of pandemic flu, dirty bombs, and smallpox. Whether one believes that greenhouse gas emissions are responsible for an increase in the frequency and severity of hurricanes, or that television and the Internet make us all too aware of potential dangers, or that the sheer magnitude of historical events such as the European Black Death of the fourteenth century, the 1556 earthquake in Shansi, China, or the Asian flu pandemic of 1919 overshadow our modern disasters by orders of magnitude, the sheer complexity and interdependencies of modern society clearly make us enormously vulnerable, whether it be to natural disasters or to terrorist attacks. The modern systems that we require to sustain our way of life—the systems that transport our energy, create our food supply, allow us to communicate over vast distances, and maintain our low infant mortality and high life expectancy—are all vulnerable to degrees that would have been unimaginable a few decades ago. Furthermore, the dollar toll from these events is increasing due to population growth in disaster-prone areas, especially in those areas susceptible to hurricanes, floods, and earthquakes.

In this new world of the twenty-first century it is essential that we anticipate such events and their potential impacts. It is impossible to know exactly what form they will take, how severe they will be, or where and when they will occur, but the value of planning has been amply demonstrated. This report is about the value of a specific area of planning and about how the United States might make improvements in that specific area. Geospatial data and tools are currently used for emergency response, but recent events have demonstrated the many ways in which our geospatial data and tools and the use we make of them fail us, both in preparing for unpredictable events and in responding to them afterwards. This report examines the current use of geospatial data and tools in emergency management and makes recommendations to improve that use.

The National Research Council's (NRC's) Committee on Geography, now the Geographical Sciences Committee, first discussed the need for this study in 2000, well before the events of September 11, 2001. Those and subsequent events led to a greater sense of urgency, a search for sponsorship, refinement of the study's charge, and to the eventual formation of a study committee in 2004 under the auspices of the NRC Mapping Science Committee. We thank the sponsors, the National Aeronautics and Space Administration, the National Oceanic and Atmospheric Administration, the National Geospatial-Intelligence Agency, and the U.S. Geological Survey, for providing funding for this study.

The committee was composed of 13 members and included scientists, social scientists, and engineers from academia, industry, government, and nongovernmental organizations. Committee members included people with experience in designing decision support tools; users of these tools; and experts in natural hazards, risk analysis, transportation, utility infrastructure, geospatial data and remote sensing, disaster planning and response, and computer and information science. The committee included members with extensive field experience in emergency management and response.

Several meetings were held to gather evidence from individuals and representatives of organizations and agencies, including emergency response practitioners and experts in geospatial data and tools. The primary information-gathering event was a workshop held on October 5-6, 2005, which included five discussion panels with approximately 25 panelists from the relevant academic disciplines and agencies and from the commercial software and data products industry. The workshop included a mix of discussion panels and breakout discussions.

This report presents the committee's findings and recommendations. It is designed to be read by any public official who is concerned to make his or her community disaster resilient: leaders of emergency response and emergency operations agencies, elected officials and citizens who are concerned about community vulnerability, agency staff who make or recommend decisions about the allocation or acquisition of resources, developers of technologies, or members of committees charged with developing policies.

Contents

SUMMARY 1

1 INTRODUCTION 9
   1.1 Scope, 9
   1.2 Statement of Task and Approach, 10
   1.3 Terms and Definitions, 12

2 THINKING ABOUT WORST CASES: REAL AND
   HYPOTHETICAL EXAMPLES 25
   2.1 The September 11, 2001, Attack on the World Trade Center, 26
   2.2 A Hypothetical Category 3 Hurricane Making Landfall
       on Long Island, New York, 30
   2.3 A Hypothetical Southern California Earthquake, 39
   2.4 Summary, 44

3 EMERGENCY MANAGEMENT FRAMEWORK 47
   3.1 The Context of Disasters, 48
   3.2 Relevant Actors, 56
   3.3 Federal Policy Relevant to Geospatial Requirements, 70
   3.4 Geospatial Data Needs, 78
   3.5 Conclusion, 79

4	THE CHALLENGE: PROVIDING GEOSPATIAL DATA, TOOLS, AND INFORMATION WHERE AND WHEN THEY ARE NEEDED	87
	4.1 Focus on Collaboration, 91	
	4.2 Geospatial Data Accessibility, 95	
	4.3 Geospatial Data Security, 105	
	4.4 Overhead Imaging, 108	
	4.5 Communication of Reports to and from the Field, 113	
	4.6 Backup, Redundancy, and Archiving, 116	
	4.7 Tools for Data Exploitation, 118	
	4.8 Education, Training, and Accessing Human Resources, 123	
	4.9 Funding Issues, 128	
5	GUIDELINES FOR GEOSPATIAL PREPAREDNESS	133
	5.1 Introduction, 133	
	5.2 Critical Elements in Successful Planning and Response, 134	
6	CONCLUDING COMMENTS: LOOKING TO THE FUTURE	145
REFERENCES		149

APPENDIXES

A	List of Acronyms	155
B	Sample Confidentiality Agreement	159
C	Preparedness Checklist	163
D	Workshop Agenda and Participants	173
E	Biographical Sketches of Committee Members and Staff	177

Summary

In the past few years the United States has experienced a series of disasters that have severely taxed and, in many cases, overwhelmed the capacity of responding agencies. Hurricane Katrina in 2005 provided perhaps the most obvious instance as millions around the world watched a region of the world's most powerful nation apparently degenerate into chaos. With modern technologies such as satellite imaging and services such as Google Earth, it was possible for anyone with access to the Internet to see the magnitude of the disaster and to marvel at how breakdown could be so complete and pervasive in an era of such technological and information abundance.

This study is about one type of information technology and the role it plays in emergency management. Geospatial data describe the locations of things on the Earth's surface, and geospatial tools manipulate such data to create useful products. Thus, this report is about the maps that are an essential part of search-and-rescue operations, about the GPS (Global Positioning System) receivers that allow first responders to locate damaged buildings or injured residents, about images that are captured from aircraft to provide the first comprehensive picture of an event's impact, about road maps that form the basis of evacuation planning, and about all of the other information connected to a location that can be used in emergency management.

Great strides have been made in the past four decades in the development of geospatial data and tools, and the Google Earth service is just one example of the power and sophistication of this type of technology. Yet no amount of technological sophistication will be sufficient to address the kinds of breakdowns that occurred in the supply and use of geospatial data and tools in recent disasters. The effectiveness of any technology is as much about the human systems in which it is embedded as about the technology itself. The committee concluded that issues of training, coordination among agencies, sharing of data and tools, planning and preparedness, and the attention and resources invested in technology turn out to be the critical factors and the ones that have to be addressed if future responses are to be more effective.

The goal of this study was to evaluate the current use of geospatial data and tools in emergency management and to make recommendations to improve that use. The study tasks assigned to the committee addressed both planning and response; the status of tools for predicting and mapping vulnerability; the types of data required for emergency management; the techniques available for discovering and accessing data from diverse sources; training requirements; and issues of data security. The committee approached the task by holding a series of meetings at which it heard evidence from individuals and representatives of organizations; organizing a workshop that included extensive formal and informal discussion; and drawing on the considerable experience of its members.

The committee's central conclusion is that geospatial data and tools should be an essential part of all aspects of emergency management—from planning for future events, through response and recovery, to the mitigation of future events. Yet they are rarely recognized as such, because society consistently fails to invest sufficiently in preparing for future events, however inevitable they may be. Moreover, the overwhelming concern in the immediate aftermath of an event is for food, shelter, and the saving of lives. It is widely acknowledged that maps are essential in the earliest stages of search and rescue, that evacuation planning is important, and that overhead images provide the best early source of information on damage; yet the necessary investments in resources, training, and coordination are rarely given sufficient priority either by the general public or by society's leaders.

In all aspects of emergency management, geospatial data and tools have the potential to contribute to the saving of lives, the limitation of damage, and the reduction in the costs to society of dealing with emergencies. Responders who know where impacts are greatest, where critical assets are stored, or where infrastructure is likely to be damaged are able to act more quickly, especially during the "golden hour" immediately after the event when there is the greatest possibility of saving lives. Geospatial data that are collected and distributed rapidly in the form of useful products allow response to proceed without the confusion that often occurs in the absence of critically important information. Indeed, it is impossible to imagine the chaos that would result if first responders were entirely unfamiliar with an area and had none of the geospatial information—maps, GPS coordinates, images—that is so essential to effective emergency management.

Massive investments have been made in geospatial data and tools over the past few decades in many areas of human activity, but the special and specific needs of emergency management—rapid operational capability and access to data, extensive planning, training of first responders, and tools that work under the difficult circumstances of search and rescue—have rarely been addressed. The committee found that while enormous amounts of data relevant and indeed essential to emergency management exist, they are frequently scattered among multiple jurisdictions, in disparate and often incompatible formats. Numerous impediments exist to data sharing, including lack of interoperability at many levels, lack of knowledge about what data exist and where, restrictions on use, lack of training on the part of users, concerns about data security, and lack of operational infrastructure in the immediate aftermath of disaster.

This report makes 12 recommendations. The first reflects the committee's central conclusion and urges that the role of geospatial data and tools should be recognized in relevant emergency management policy documents, directives, and procedures (Chapter 4):

> **RECOMMENDATION 1: The role of geospatial data and tools should be addressed explicitly by the responsible agency in strategic planning documents at all levels, including the National Response Plan, the National Incident Management System, the Target Capabilities List, and other pertinent plans, procedures, and policies (including future Homeland Security Presidential Directives). Geospatial procedures and plans developed for all but the smallest of emergencies should be multiagency, involving all local, state, and federal agencies and nongovernmental organizations (NGOs) that might participate in such events.**

In the early 1990s a new effort to coordinate the production, distribution, and use of geospatial data began under the rubric of the National Spatial Data Infrastructure (NSDI). Standards have been developed and implemented, clearinghouses have been built and access portals deployed, and today the NSDI provides a coherent framework for the sharing of geospatial data. To date, however, the special needs of emergency management have not been recognized as fully as the committee considers desirable, and emergency management is only weakly represented within the NSDI's existing governance structure. Accordingly, the committee's second and third recommendations seek to strengthen the NSDI as a framework for the effective sharing of geospatial data for emergency management. Since the Department of Homeland Security has been given responsibilities for geospatial data coordination for emergency management, the committee specifically identifies it in the recommendations, proposing that it play a leading role in strengthening the NSDI in this way (Sections 4.1 and 4.2):

> **RECOMMENDATION 2: The current system of governance of the NSDI should be strengthened to include the full range of agencies, governments, and sectors that share geospatial data and tools, in order to provide strong national leadership. The Department of Homeland Security (DHS) should play a leading role in ensuring that the special needs of emergency management for effective data sharing and collaboration are recognized as an important area of emphasis for this new governance structure.**

> **RECOMMENDATION 3: A new effort should be established, within the framework of the NSDI and its governance structure and led by DHS, to develop policies and guidelines that address the sharing of geospatial data in support of all phases of emergency management. These policies and guidelines should define the conditions under which each type of data should be shared, the roles and responsibilities of each participating organization, data quality requirements, and the interoperability requirements that should be implemented to facilitate sharing.**

Security is one of the many reasons cited by organizations for failing to share data and failing to make data available in support of emergency response. The committee's fourth recommendation seeks to address this issue through a system that would restrict access where necessary to appropriately authorized emergency management personnel (Section 4.3):

> **RECOMMENDATION 4: DHS should lead, within the framework of the NSDI, the development of a nationally coordinated set of security requirements for data to be shared for emergency preparedness and response. All organizations should implement these guidelines for all data shared in support of emergency management and should use them where necessary to restrict access to appropriately authorized personnel. In concert with these efforts, the leveraging of existing organizations that could potentially serve as a "clearinghouse" for critical infrastructure data should be explored.**

Section 4.4 of this report describes the problems that occur in the immediate aftermath of events when geospatial data must be acquired as quickly as possible to assess impacts and plan response and recovery. All too often a lack of planning produces delays as organizations scramble to overcome administrative roadblocks:

> **RECOMMENDATION 5:** Standing contracts and other procurement mechanisms should be put in place at local, regional, and national levels by the responsible agencies to permit state and local emergency managers to acquire overhead imagery and other types of event-related geospatial data rapidly during disasters.

Hurricane Katrina and other recent events have shown all too clearly the potential magnitude of disasters and their ability to overwhelm agency resources. Chapter 2 of this report describes the experience of the attacks on the World Trade Center as well as two additional hypothetical scenarios, one a major storm in the New York area and the other a major earthquake in the Los Angeles Basin. The committee believes that events of this magnitude should be the basis for extensive preparedness exercises, since they will allow many of the issues that arose during recent responses to be anticipated and explored (Section 4.5):

> **RECOMMENDATION 6:** Interpersonal, institutional, technical, and procedural communications problems that currently inhibit communication between first responders in the field and emergency operations centers, emergency management agency headquarters, and other coordinating centers should be addressed through intensive preparedness exercises by groups involved in all aspects of disaster management. Such exercises should be tailored to focus on clear objectives with respect to the use of geospatial data and assets. They should involve decision-making representatives from all levels of government, as well as other relevant organizations and institutions, and should be coordinated nationally so that common problems can be identified. They should be realistic in their complexity and should allow participants to work carefully through the geospatial challenges posed by disasters, including the difficulty of specifying requirements, the difficulty of communicating in a context of compromised infrastructure, and the difficulty of overcoming logistical obstacles.

It is surprising perhaps that despite the intensity of the efforts that went into recovering from the World Trade Center attacks, very little documentation exists detailing the geospatial data and tools that were employed that might serve as a basis for improved responses to similar events in the future and as a basis for training. The experience of recent events, particularly the World Trade Center attacks, also points to the need for effective policies regarding the backing up of geospatial data and tools in geographically separate locations (Section 4.6):

> **RECOMMENDATION 7:** DHS should revise Emergency Support Function 5 of the National Response Plan to include backup and archiving of geospatial data, tools, and procedures developed as part of disaster response and recovery. It should assign responsibility for archiving and backup in the Joint Field Offices during an incident to the Federal Emergency Management Agency, with an appropriate level of funding provided to perform this function.

As noted earlier, geospatial data and tools are now widely deployed in many areas of human activity. Emergency management presents specific circumstances, however, and demands a different approach to the development and deployment of technologies. The committee finds that there is a significant gap between the needs of emergency management and the capabilities of current systems and recommends (Section 4.7) the following:

> **RECOMMENDATION 8:** The National Science Foundation and federal agencies with responsibility for funding research on emergency management should support the adaptation, development, and improvement of geospatial tools for the specific conditions and requirements of all phases of emergency management.

The committee believes strongly that if geospatial technologies are to become an integral part of emergency response and recovery, they must be part of the day-to-day operations of emergency managers and responders at all levels of government and there must be an increase in the number of personnel trained in the use of geospatial data and tools available to support emergency management. The committee's next three recommendations address this issue (Section 4.8):

> **RECOMMENDATION 9:** Academic institutions offering emergency management curricula should increase the emphasis given to geospatial data and tools in their programs. Geospatial professionals who are likely to be involved in emergency response should receive increased training in emergency management business processes and practices.

> **RECOMMENDATION 10:** The Federal Emergency Management Agency should expand its team of permanent geospatial professionals and develop strategies that will lead to their more rapid deployment both in response to events and in advance of events when specific and reliable warnings are given.

> **RECOMMENDATION 11:** The Department of Homeland Security should establish and maintain a secure list of appropriately quali-

fied geospatial professionals who can support emergency response during disasters.

Finally, the committee found that funding for geospatial preparedness is insufficient and the funding that exists is often used ineffectively (Section 4.9):

> **RECOMMENDATION 12:** To address the current shortfall in funding for geospatial preparedness, especially at the state and local levels, the committee recommends: (1) DHS should expand and focus a specifically designated component of its grant programs to promote geospatial preparedness through development, acquisition, sharing, and use of standard-based geospatial information and technology; (2) states should include geospatial preparedness in their planning for homeland security; and (3) DHS, working with the Office of Management and Budget, should identify and request additional appropriations and identify areas where state, local, and federal funding can be better aligned to increase the nation's level of geospatial preparedness.

Besides these recommendations, the report also provides a set of more detailed guidelines for the assessment of geospatial preparedness in emergency management organizations in Chapter 5 and Appendix C. The list is not intended to be exhaustive, but rather to provide a basis for enhancing geospatial preparedness and for directing planning and investment.

In essence, the report paints a picture of technological abundance, but of geospatial data and tools that despite their power have not yet been applied systematically and appropriately to emergency management. It lists numerous institutional factors that have inhibited the effective deployment of technology and numerous reasons why organizations have failed to anticipate and plan for the particular circumstances created by disasters. The committee hopes that the recommendations made in this report, and the examples and guidelines that it provides, will help to create a world in which future responses to disasters will be faster and more effective.

1

Introduction

1.1 SCOPE

We live in an age of technological abundance. Computers, the Internet, satellites, and many other tools provide us with unprecedented ability to collect, store, analyze, and distribute information on all aspects of the planet we inhabit, the communities in which we live, and the daily activities that we perform. We have access to high-resolution images of our neighborhoods; we can obtain driving directions through cell phones; we can track the movement of pets, people on probation, and vehicles with Global Positioning System (GPS) tracking devices; and we can plan new developments using geographic information systems (GIS). In all of these examples the data and tools exploit our ability to know *where* events, activities, individuals, streets, or buildings are to be found on the Earth's surface. In other words, they include information that might be presented in the form of a map. Today, we term such data and tools *geospatial* to distinguish them from other types of data and tools, and they are essential in virtually all aspects of human activity, from the operations of government agencies and private corporations to the daily lives of the general public. They are encountered in the form of paper maps, in-car navigation systems, Internet sites, the software and databases of local government and utility companies, the sophisticated analytical, mapping, and visualization tools that support decision making in many private- and public-sector organizations, and many other forms. Because of their usefulness for so many diverse applications—from zoning and taxation, to environmental management, to the national census, and of course, emergency response—communities at all levels have made massive investments in acquiring geospatial information, converting it to digital form, and maintaining computer-based systems for accessing and using it. Often, multiple applications need the same geospatial data sets, allowing for efficiencies or redundancies in data development, depending on the amount of coordination between organizations.

This report is about the geospatial data and tools that are available for one particular application, that of preparing for and responding to emergencies. It discusses how those resources are utilized and the impediments that may exist to their greater and more effective utilization. Although in testimony the committee was told that "successful emergency response starts with a map," the experience of recent disasters such as Hurricane Katrina and the attacks of September 11, 2001, has shown that the geospatial data and tools that exist within our communities have not been integrated effectively into disaster planning, response, and recovery (Sidebar 1.1). There are many reasons for this, and they are explored in this report. The committee also examines the consequences of underutilization, which are often disastrous, in the form of loss of life, damage to property, and damage to the environment. The report's recommendations point to steps that can be taken to address this serious issue at local, national, and international levels through increased utilization and more effective integration of geospatial data and tools into emergency management processes. As MacFarlane (2005, p. 124) notes in a report on the use of geographic information systems in emergency management in the United Kingdom, "The principles, both technical and operational, . . . are established . . . and the technical enablers are all proven. It will now require vision and leadership to realize the gains." Closer to home, a recent report from the National Governors Association concluded that "despite the promise of GIS technology for strengthening homeland security and its growing popularity across government, its use is not yet ubiquitous. To be useful during an emergency, the mapping tools and underlying data must be in place before the event occurs. While an emergency operations center (EOC) would require the tools and data other government agencies use every day, many states still lack an organizational and operational connection between the EOC and those other agencies. Often executive leadership is necessary to make these connections" (National Governors Association, 2006, p. 7).

1.2 STATEMENT OF TASK AND APPROACH

The committee was charged with assessing the status of the use of geospatial data, tools, and infrastructure in disaster management and making recommendations to increase and improve that use. Specifically, the study tasks were to

1. Assess the value of geospatial data and tools in disaster planning and disaster response;
2. Identify the status of and needs for decision support tools that assimilate model predictions and data for mapping vulnerability to catastrophe, scenario testing, disaster planning, and logistical support;
3. Identify the mission-critical data requirements for effective decision making;
4. Examine technical and institutional mechanisms that enable rapid discovery, access, and assemblage of data from diverse sources;
5. Assess training needs for developers and users of spatial decision support systems; and
6. Examine potential conflicts between issues of security and the need for open access to data.

The committee met four times to gather the information needed for the study and write the report. The first two meetings included presentations by various federal agencies and private industry organizations that described how geospatial data and tools are currently being used for emergency management and discussed issues related to their use. The third meeting was a workshop consisting of five discussion panels. The 25 panelists included a broad range of specialists in various aspects of emergency management from city, county, state, and federal government, private industry, academia, nongovernmental organizations, and the United Nations. (See Appendix D.) Breakout sessions with these experts were also held to obtain further insights on the study tasks. After acquiring and synthesizing a general overview of the value of and needs for geospatial data and tools in emergency management through these meetings, published documents, and the expertise of its members, the committee focused on the status of their *use* (i.e., how much and how effectively they are currently being used and what is preventing better utilization). Recommendations were then developed to address each of these challenges.

The report is structured as follows. The remainder of this chapter defines the major terms used in the report. Chapter 2 presents three scenarios, one real and two hypothetical, to illustrate how geospatial data and tools are currently used in emergency response and how better utilization could improve response. Chapter 3 goes into more detail about disasters and emergency response, to provide context and describe the needs of emergency responders, and reviews how current federal-level emergency management policies address geospatial data and tools. Chapter 4 then presents issues and challenges that are impeding the effective use of geospatial data and tools and provides recommendations for addressing these challenges. Chapter 5 provides guidelines that can be used by emergency managers to review their "geospatial preparedness" to respond to disasters. Finally, Chapter 6 offers some thoughts for the future.

> **Sidebar 1.1**
> **Google Earth**
>
> Google Earth,[a] an Internet-based service originally developed by Keyhole, became instantly popular when it was rebranded and released by Google in early 2005. It allows users to view the Earth as a whole, zooming from global to local scales, using high-resolution imagery that shows individual buildings and vehicles, and to simulate a magic carpet ride over any part of the Earth's surface. By releasing an application programmer interface (API), Google enabled thousands of individuals to add their own data and their own applications, and to make them easily accessible to anyone. In many ways, Google Earth represents a dramatic improvement in the accessibility of geospatial data and tools, allowing the general public to explore the Earth's surface in ways that had previously been available only to geospatial professionals.
>
> In the immediate aftermath of Hurricane Katrina, high-resolution images began to appear on the Google Earth site, showing in detail the impacts of the disaster. People from all over the world could explore the impacted area, seeing the levee breaks in New Orleans, the extent of the flooding, the damage to buildings, and the impacts on the environment. Images from Google Earth appeared on television newscasts around the world. Within the impacted area, however, where computers were damaged, electrical power networks were destroyed, and Internet communications were disrupted, it was impossible for emergency managers to make use of Google Earth's data and tools for days and in some cases weeks or months. Paradoxically, access to geospatial data and tools resembled a donut—abundant far away from the impact area, but almost nonexistent where it was most needed in the donut's center.
>
> ---
>
> [a]*http://earth.google.com.*

1.3 TERMS AND DEFINITIONS

1.3.1 Emergency Management

In this report, the term *emergency* is used to mean a sudden, unpredictable event that poses a substantial threat to life or property. Emergencies vary in magnitude, depending on the degree of threat, and they also vary in duration and in the geographic extent of their impacts. A *disaster*

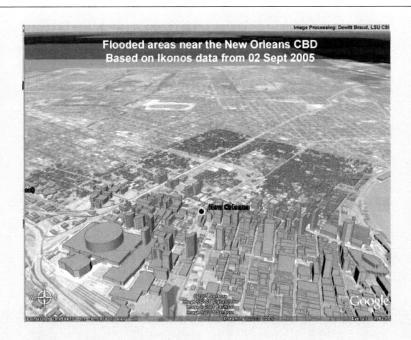

Google Earth screen capture showing oblique high-resolution view of New Orleans central business district in foreground depicting post-Katrina flooding. From Andrew Curtis, Louisiana State University (LSU), a PowerPoint presentation on January 25, 2006, at the National Centre for Geocomputation (*http://ncg.nuim.ie/ncg/events/20060125/*). Image courtesy LSU World Health Organization Collaborating Center for Remote Sensing and GIS for Public Health (WHOCC) Laboratory. Image processing of IKONOS satellite imagery by DeWitt Braud, LSU Coastal Studies Institute. Google Earth KMZ file and image produced by Jason K. Blackburn, Ph.D., LSU WHOCC. Image courtesy Google Earth™ mapping service, used with permission.

is defined as a calamitous event that overwhelms the impacted community's ability to respond effectively. *Catastrophes* are very large disasters that often require concerted national and international response efforts. All three terms were used in the committee's Statement of Task, but to avoid confusion only *emergency* and *disaster* are used in this report, the latter when it is desirable to emphasize severity or magnitude and the tendency of such events to overwhelm communities. The terms *event* and *incident* are used interchangeably throughout the report.

Disasters suddenly result in extensive negative economic and social consequences for the populations they affect, typically including physical injury, loss of life, property damage, physical and emotional hardship, destruction of physical infrastructure, and failure of administrative and operational systems. All disasters threaten the general welfare of some populace; thus, government intervention is warranted to minimize the negative consequences of disaster and, ultimately, to restore order. Disasters are often classified by cause (Alexander, 2000; Burton et al., 1993; Cutter, 2001): natural (e.g., floods, droughts, landslides, volcanoes, hurricanes, earthquakes, winter storms, tsunami), technological (e.g., chemical spills or releases, computer failures, train derailments, plane crashes, power outages, bridge collapses), or social (e.g., riots, willful acts such as arson or terrorism). However contemporary conceptual frameworks focus on the common elements of disaster incidents regardless of type or cause, referring to the full breadth of incidents as *"all-hazard."*

Emergency management is the organization and management of resources and responsibilities for dealing with all aspects of emergencies. Four phases of emergency management are generally recognized: *preparedness, response, recovery,* and *mitigation*. Emergency management involves plans, structures, and arrangements established to engage the normal endeavors of government and voluntary and private agencies in a comprehensive and coordinated way to respond to the whole spectrum of emergency needs.

An *emergency operations center* is a facility established to serve as a focus for response and recovery support. EOCs vary dramatically in their configuration and purpose. They may house key emergency management staff and also provide the main access point for geospatial data and tools, producing maps and other hard-copy geospatial products for distribution to teams in the field. Other EOCs may contain only liaisons that help with coordination and provision of resources, and in such cases, the locus of decision making may be elsewhere—for example, at the Joint Field Office for federal assets and at various command posts for local assets.

Chapter 3 provides a more comprehensive discussion of all aspects of emergency management and describes the importance of geospatial data and tools in each aspect.

1.3.2 Geospatial Data and Tools

All disasters have a temporal and geographic *footprint* that identifies the duration of impact and its extent on the Earth's surface. The term *geospatial* is used to refer to those interdependent resources—imagery, maps, data sets, tools, and procedures—that tie every event, feature, or entity to a location on the Earth's surface and use this information for some purpose. Location must be expressed in some standard and readily understood form, such as latitude-longitude, street address, or position in some coordinate system. GPS is today a very cost-effective way of associating an event, feature, or entity with a location and, thus, of making data geospatial. Consistent use of such associations across a range of data sets makes possible their integration for a variety of purposes, including display as maps and use for analysis and modeling.

Although location is an essential part of any item of geospatial data, it is the ability to link a location to the properties of events, features, or entities at that location that gives geospatial data their value. (To be consistent with practice among geospatial professionals, the term *feature* is used throughout this report to refer to any event, feature, or entity whose location and attributes are recorded in a geospatial data set.) These properties are collectively termed *attributes* and may include the owner of a parcel of land, the population of a neighborhood, the temperature of a point in a burning building, or the wind speed and direction at a point in a hazardous plume.

The terminology of geospatial data and tools is highly specialized, and efforts have been made throughout this report to minimize the use of technical terms and, where appropriate, to clarify their meaning. Particular resources and capabilities are referenced at points, both to serve as examples and to provide additional insight for the more technically proficient reader. Those who wish to explore this field in more depth should consult one of the many introductory texts, such as those by Clarke (2003), DeMers (2005), Longley et al. (2005), and Worboys and Duckham (2004).

Throughout this document, three types of geospatial data are discussed: framework data, foundation data, and event-related data. *Framework* data comprise the seven geographic themes that are most commonly produced and used by most organizations in their day-to-day geospatial activities,[1] and which potentially provide a set of landmarks on the Earth's surface to which other data can be tied (for example, if an event occurs at a street intersection and the location of that street intersection is known in the framework, then the location of the event is also known). The Federal Geographic Data Committee (FGDC) identifies these seven themes as geodetic control, orthoimagery, elevation, transportation, hydrography, governmental units, and cadastral information, since all of these are used in various contexts as frameworks within which other features can be located (see Sidebar 1.2). Although they fall outside the range of data themes in framework data, *foundation* data are also routinely col-

---

[1]*http://www.fgdc.gov/framework/.*

> **Sidebar 1.2**
> **Framework Data Layers**
>
> The Federal Geographic Data Committee has identified seven themes as forming the geospatial data framework:
>
> 1. Geodetic control: the very accurate system of measurements used to establish the shape of the Earth and to lay out its basic coordinate systems
> 2. Orthoimagery: high-resolution images derived from aerial photographs or satellites and corrected geometrically as if every location were vertically below the eye
> 3. Elevation: data on the elevation of the Earth's surface at densely sampled locations
> 4. Transportation: the locations and properties of streets, roads, railroads, and other transportation features
> 5. Hydrography: the locations and properties of rivers, lakes, and coastlines
> 6. Governmental units: the locations and properties of administrative areas such as states, counties, and municipalities
> 7. Cadastral: the map of land ownership, showing the locations of property boundaries
>
> SOURCE: *http://www.fgdc.gov/framework/*.

lected to support the day-to-day operations of private- or public-sector organizations or agencies. Foundation data themes typically relate to a specific organization's mandate and thus complement the framework data themes. Examples of foundation data include maps or data sets of soils, land use, weather, underground pipes, or overhead power lines. Given the diversity of such themes, foundation data are a valuable resource for emergency management and will reflect the state of an area prior to a disaster event, providing an essential baseline. Finally, *event-related* data include all those items collected specifically to respond to and recover from a particular disaster event. Such data include the locations of casualties, the locations of response resources, and imagery and inventories of property and environmental damage. They might also include data gathered in real time from sensors monitoring event-related phenomena such as earthquake aftershocks or chemical plumes, since such data can be important in managing response.

Throughout this report, the committee differentiates between *data* and *information*. The latter term connotes usefulness for some purpose, as when

field observations of damage are checked and compiled into maps that are designed to be used by first responders, or when raw measurements of atmospheric conditions are converted into predictions of hurricane tracks. Geospatial information is typically what is produced when geospatial tools and procedures are applied to geospatial data.

To make effective use of a geospatial data resource, a user needs access to that data resource's *metadata*. Simply, metadata are "data about data" and describe the content, quality, condition, level of geographic detail, and other characteristics of geospatial data resources,[2] whether these are images, paper maps, or digital data sets (see Sidebar 1.3 and Figure 1.1).

Metadata allow one person to describe a geospatial data set to another person, allow users to search for geospatial data sets within on-line catalogs, allow various kinds of automated processing of geospatial data, and allow producers of geospatial data to track and manage the production process. Metadata help to ensure that data are used appropriately; without metadata, data may be improperly applied, resulting in inappropriate conclusions or decisions. Thus, metadata play a critical role in all aspects of geospatial data use. When agencies have agreed to share data resources to provide an integrated resource for emergency management, it is essential that accurate metadata exist for each resource.

Remote sensing describes the collection of data from a wide range of automated systems, including satellites and aircraft equipped with imaging sensors, ground-based sensors for detecting biological and chemical agents, and ground-based surveillance cameras. Imaging sensors may be passive, relying on radiation reflected or emitted from the scene, or active, emitting signals and detecting their echoes to build images of three-dimensional structures. Many such systems have important applications in emergency management and are cited at various points in this report.

Geographic information systems are software systems used to capture, store, manage, analyze, and display geospatial data resources (see Figure 1.2). The "geographic" element of the name refers to the use of location, specifically a coordinate system, as an organizing principle for these data resources. Essentially, GIS integrate numerous functions that can be applied to geospatial data into a single, integrated tool set, just as Microsoft Word integrates numerous functions that can be applied to the creation and editing of text. GIS are among the most important and widely used of geospatial tools, and their functions allow emergency managers to integrate geospatial data, create maps, produce statistical summaries, and perform many other essential functions.

---

[2]*http://www.fgdc.gov/metadata/metadata.html.*

> **Sidebar 1.3**
> **Example of Metadata**
>
> **High-Resolution Shoreline Data (2007)**
>
> ***Identification Information:***
> ***Citation:***
> ***Citation Information:***
> ***Originator:*** National Oceanic and Atmospheric Administration (NOAA), National Ocean Service (NOS), National Geodetic Survey (NGS)
> ***Publication Date:*** 2007
> ***Title:*** High-Resolution Shoreline Data (2007)
> ***Geospatial Data Presentation Form:*** vector digital data
> ***Publication Information:***
> ***Publication Place:*** Silver Spring, Md.
> ***Publisher:*** NOAA's Ocean Service, National Geodetic Survey (NGS)
> ***Online Linkage:*** http://www.ngs.noaa.gov/RSD/shoredata/ NGS_Shoreline_Products.htm
> ***Description:***
> ***Abstract:*** These data provide an accurate high-resolution shoreline compiled from remote-sensing data. The vector shoreline data may be suitable as a geographic information system (GIS) data layer. This metadata describes information for both the line and point shapefiles. The NGS attribution scheme Coastal Cartographic Object Attribute Source Table (C-Coast) was developed to conform the attribution of various sources of shoreline data into one attribution catalog. C-COAST is not a recognized standard but was influenced by the International Hydrographic Organization's S-57 Object-Attribute standard so that the data would be more accurately trans-

Computer-assisted design (CAD) systems are widely used to create and manage three-dimensional digital models of buildings and other engineering structures. They have proven invaluable in modeling the collapse of structures such as the twin towers of the World Trade Center in order to provide direction for rescue efforts. When accurately registered to the Earth's surface, CAD data can be usefully combined with other geospatial data.

Many web sites now provide access to large collections of geospatial data sets, which can be discovered, assessed, and possibly downloaded by remotely located users. Such sites are known as *clearinghouses*, *geolibraries*, *archives*, or *geoportals*. The last term reflects the most recent advances in this field, since a geoportal provides a single point of entry to

INTRODUCTION 19

> lated into S-57. When complete, the data will be made available on-line: http://www.ngs.noaa.gov/RSD/shoredata/NGS_Shoreline_Products.htm.
> **Purpose:** The shoreline and associated data layers were originally intended to support the NOAA nautical chart production. These data sets have been cleaned and reformatted to support the efforts of supplying accurate shoreline data layers for a coastal GIS database. These data sets may be beneficial for performing change analysis for erosion and accretion studies, land-use planning, determination of boundary extent, and other types of decision making.
> **Time Period of Content:**
> **Time Period Information:**
> **Range of Dates/Times:**
> **Beginning Date:** 20061001
> **Ending Date:** 20070930
> **Currentness Reference:** publication date
> **Status:**
> **Progress:** planned
> **Maintenance and Update Frequency:** As needed
> **Spatial Domain:**
> **Bounding Coordinates:**
> **West Bounding Coordinate:** 141
> **East Bounding Coordinate:** −61.1
> **North Bounding Coordinate:** 74.8
> **South Bounding Coordinate:** −14.6
>
> SOURCE: NOAA Data Explorer. Available at *http://oceanservice.noaa.gov/dataexplorer/welcome.html* [accessed on October 24, 2006].

resources that may be located in many different repositories. The Geospatial One-Stop[3] (GOS) is a geoportal sponsored as part of the administration's E-Government Initiative and designed to provide a single point of entry to geospatial data and web-based tools (Figure 1.3).

Spatial decision support systems (SDSS) are designed to provide the essential information needed by decision makers when those decisions involve location. For example, an SDSS might be used to design evacuation routes, to select optimum locations for response teams, or to allocate evacuees to shelters. SDSS are in effect specialized GIS, designed to be

---

[3] *http://gos2.geodata.gov/wps/portal/gos/*.

20

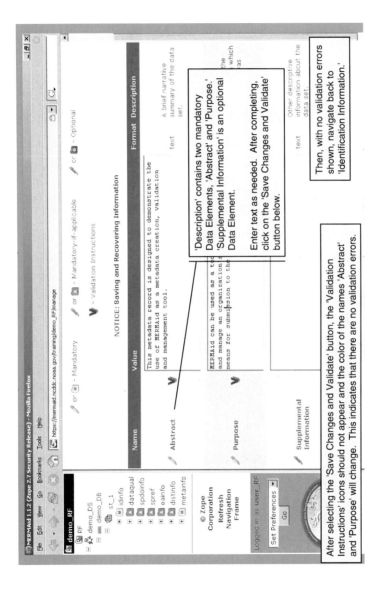

FIGURE 1.1 Example of the metadata development tool called Metadata Enterprise Resource Management Aid (MERMAid) by the National Oceanic and Atmospheric Administration. SOURCE: *http://www.ncddc.noaa.gov/Metadata/docs/pdfv112* [accessed on October 24, 2006].

INTRODUCTION   21

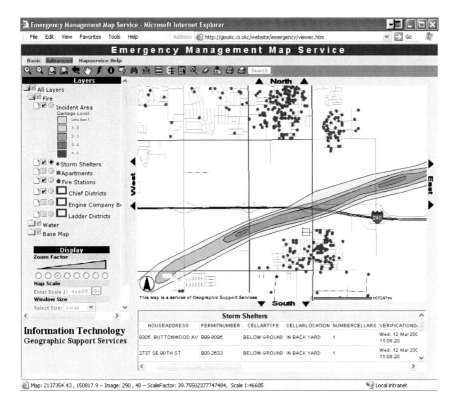

FIGURE 1.2 Example of a screen shot from a GIS that can be used for emergency management. SOURCE: Schad Meldrum, City of Oklahoma City.

used by decision makers to perform specific functions that often involve comparisons between many possible alternatives. They allow the decision maker to consolidate, summarize, model, and transform data to support such tasks as analytical reporting, visualization, and trend analysis.

Geospatial tools are software and hardware systems that perform specific operations on geospatial data. They include GIS and SDSS, as well as more limited tools and equipment designed specifically for such functions as the analysis and processing of images, the reformatting of data, or the acquisition of GPS measurements. They also include web-based tools and sites that offer limited mapping and analysis functions through the user's own web browser.

Geospatial infrastructure is the set of institutions, people and skills, standards, educational programs, and other arrangements that provide the context within which geospatial data and tools are used. In the United

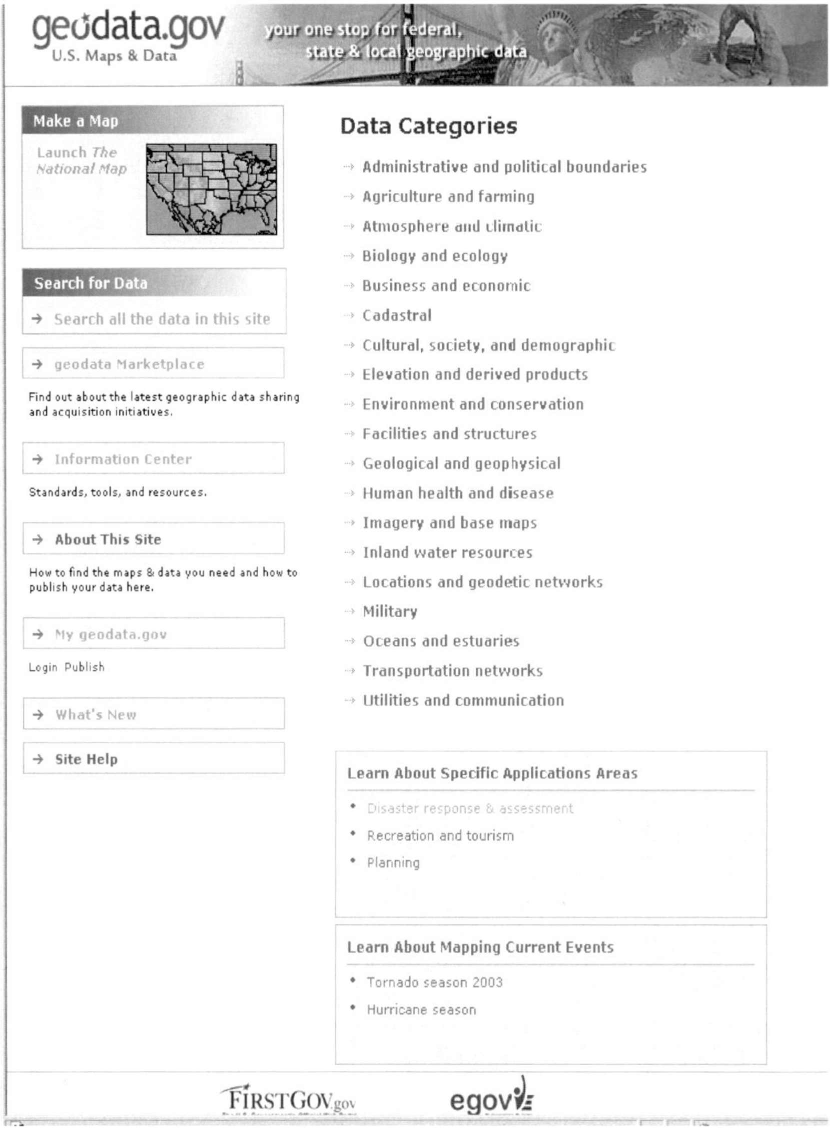

geodata.gov
U.S. Maps & Data

your one stop for federal, state & local geographic data

Make a Map
Launch *The National Map*

Search for Data
→ Search all the data in this site

→ geodata Marketplace
Find out about the latest geographic data sharing and acquisition initiatives.

→ Information Center
Standards, tools, and resources.

→ About This Site
How to find the maps & data you need and how to publish your data here.

→ My geodata.gov
Login Publish

→ What's New

→ Site Help

Data Categories
→ Administrative and political boundaries
→ Agriculture and farming
→ Atmosphere and climatic
→ Biology and ecology
→ Business and economic
→ Cadastral
→ Cultural, society, and demographic
→ Elevation and derived products
→ Environment and conservation
→ Facilities and structures
→ Geological and geophysical
→ Human health and disease
→ Imagery and base maps
→ Inland water resources
→ Locations and geodetic networks
→ Military
→ Oceans and estuaries
→ Transportation networks
→ Utilities and communication

Learn About Specific Applications Areas
* Disaster response & assessment
* Recreation and tourism
* Planning

Learn About Mapping Current Events
* Tornado season 2003
* Hurricane season

FIRSTGOV.gov   egov

FIGURE 1.3 Data categories available in the Geospatial One-Stop.

States, the *National Spatial Data Infrastructure* refers to those elements of geospatial infrastructure that were instituted beginning in the early 1990s in response to reports of the National Research Council (NRC, 1993), Executive Order 12906,[4] the Office of Management and Budget's Circular A-16,[5] and the ongoing efforts of the Federal Geographic Data Committee.[6]

Interoperability is the ability of products, tools, systems, or processes to work together to accomplish a common task. In the context of this report, the term refers specifically to geospatial data and tools. Often, the focus of interoperability is on the software systems that are used to capture, store, process, analyze, and display geospatial data, and on the data sets that must be exchanged between them. Because these systems have often been developed by different vendors using their own proprietary standards, there are frequently problems in exchanging data between them. Different agencies may also use different standards, different classification systems, or different terms to describe the same things. Thus, one way to move toward interoperability is through the application of open, vendor-neutral, nonproprietary standards that are developed in a voluntary consensus-based process.

Interoperability describes an ideal world in which problems of exchange have been addressed, allowing data and tools to be shared freely and rapidly. In a broader sense, however, interoperability must deal not only with data and tools but with differences that may exist between components of computer hardware, the networks that link them, and the communications technologies that operate on those networks.[7] In the broadest possible sense, interoperability also refers to the processes, policies, and personnel of organizations and institutions, and this broadest sense is particularly pertinent to effective emergency management.

Geospatial preparedness reflects the overall capability and capacity necessary to enable all levels of government and the private sector to assemble and utilize geospatial data resources, GIS software and hardware, and SDSS to perform essential emergency management functions in order to minimize loss of life and property.

---

[4]*http://www.archives.gov/federal-register/executive-orders/pdf/12906.pdf.*
[5]*http://www.whitehouse.gov/omb/circulars/a016/a016_rev.html.*
[6]*http://www.fgdc.gov.*
[7]*http://www.opengeospatial.org/resources/?page=faq#45/.*

2

Thinking About Worst Cases: Real and Hypothetical Examples

To better understand the benefits of the use of geospatial data and tools in emergency management, this chapter presents a series of scenarios—one real, two hypothetical—to illustrate how geospatial data and tools are or could have been used to support both preparedness and response. The chapter begins with a description of the role played by geospatial data and tools in the response to the September 11, 2001, attacks on the World Trade Center and the lessons learned from that experience. It then uses two hypothetical cases to describe geospatial preparedness as it currently exists, possible scenarios of post-event response and recovery, and the lessons that might be learned by thinking through such scenarios. These three events were chosen to exemplify the difference in experiences between slower-onset events with significant warning time, such as a hurricane, and rapid-onset events such as terrorist attacks and earthquakes that occur without warning. The settings for the events—metropolitan New York and Southern California—were selected to demonstrate the issues associated with complex urban systems with multiple jurisdictions, significant and vulnerable property and infrastructure, and large and diverse populations. While the second and third events are hypothetical, they nevertheless draw on experiences from events that have occurred in the past 15 years or so, such as Hurricane Katrina in 2005.

As befits the context, the emphasis in all three cases is on the role played by geospatial data and tools. The descriptions focus on how geospatial data and tools were or would have been used in all four phases of management: preparedness, response, recovery, and mitigation (see

Section 3.1 for a complete discussion of these stages). By examining cases such as these we can explore the value of the use of geospatial data and tools, the gap between today's typical levels of geospatial preparedness on the one hand, and what could be done to lessen the impact and consequences of disasters on the other. Although the second and third scenarios are hypothetical, they almost inevitably will become real at some date in the future.

2.1 THE SEPTEMBER 11, 2001, ATTACK ON THE WORLD TRADE CENTER

2.1.1 Background

The terrorist attack that took place on September 11, 2001, in New York City resulted in thousands of lives lost, the collapse of the twin towers of the World Trade Center as well as damage to adjacent buildings, and extensive disruption of transportation and other lifeline systems, economic activity, and social activities within the city and the surrounding area. When the final accounting takes place, this attack will almost certainly constitute one of the most deadly and costly disaster events in U.S. history. In a very real sense, the September 11th tragedy, the nature of the damage that occurred, the challenges that the city's emergency response faced, and the actions that were undertaken to meet those demands can be seen as a proxy—albeit a geographically concentrated one—for what a major earthquake can do in a complex, densely populated modern urban environment. Like an earthquake, the terrorist attack occurred with virtually no warning. As would be expected in an earthquake, fires broke out and many structures collapsed. As has been observed in major urban earthquakes and other disasters (e.g., Hurricane Katrina), structures housing facilities that perform critical emergency functions were destroyed, heavily damaged, or evacuated for life-safety reasons. Thus, the attack and its aftermath provide a useful laboratory for exploring a variety of engineering and emergency management issues and for learning lessons that can be applied in many other contexts.

The initial attack caused the collapse of the two main towers of the World Trade Center, but flaming debris from the impact of the first jet ignited a fire in a fuel tank in Building 7, weakening the structure so significantly that this building also collapsed, destroying one of the most sophisticated emergency operations centers (EOCs) in the country. Housed in the EOC were geospatial tools and municipal data that had been carefully accumulated over years to respond to numerous types of emergencies. Backup data, which were stored in another part of the same building, were also lost. Another copy of the data was stored on computers in the Department of Information Technology and Telecommunications building two blocks north of the World Trade Center, but this building was abandoned shortly after the twin towers collapsed and the data were rendered useless when dust and fumes from the collapsing buildings permeated the area and all power and communications were lost. Thus, despite extensive efforts to prepare for such events by establishing the EOC, the first maps distributed in support of the response were produced from a database stored at Hunter College, used primarily for research and teaching, which fortuitously was available for use in response.

A combination of factors culminated in an unprecedented use of geospatial data and tools. First, the devastation was beyond the imaginative capabilities of emergency management professionals—there was simply no appropriate script for such an event—and the demand for information proved to be immense. Second, Manhattan is unique in the United States because of the density of its high-rise buildings, the complexity of its infrastructure, and the value of its real estate. Because of this, and the logistical needs that follow, the City of New York had already undertaken intense mapping efforts, resulting in the production of highly accurate geospatial databases. Third, even though this was a very large disaster, the scale of events in New York was highly localized, with primary impacts concentrated in a small geographic area that could be mapped and imaged comparatively easily and quickly. These three factors combined to make both imagery and maps particularly useful.

The geospatial operations center was moved to Pier 92 on Friday, September 14th, and organized into the Emergency Mapping and Data Center (EMDC). Equipment and software were donated by numerous vendors, and New York State contracted for and provided the city with aerial imagery, detailed elevation data, and thermal data on a daily basis for six weeks. Eventually, more than 100 volunteers assisted in responding to more than 3,000 individual requests for geospatial information in support of the search and recovery efforts at the World Trade Center site. Responders were able to track mobile offices, medical support teams, heat from the fires, hazards, debris, and the daily progress of the search teams using geospatial data and tools.

2.1.2 Lessons Learned

It is clear that the primary responders, the New York Fire Department (FDNY), were able to make extensive use of maps and remote-sensing data during this disaster (see Figures 2.1, 2.2, and 2.3 for examples). Geospatial data and tools played a major role in rapidly integrating, analyzing, and visualizing data for emergency response, including the following:

• Providing aerial imagery, detailed elevation data, and thermal data on a daily basis to support damage assessment and ongoing response effort;
• Tracking daily progress of the search teams;
• Tracking mobile offices and medical support teams;
• Mapping heat from the fires, hazards, and debris; and
• Providing key information on subways, telecommunications, and other infrastructure layers as well as fuel and coolant storage tanks.

Significant lessons were learned from the World Trade Center response effort, which as noted earlier was unprecedented in many ways,

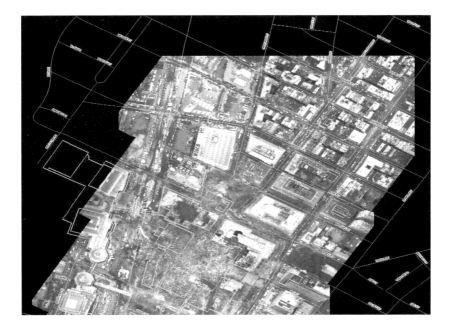

FIGURE 2.1 Aerial imagery with building footprints and street names overlaid on it. Picture provided with the permission of the New York City Office of Emergency Management.

THINKING ABOUT WORST CASES 29

FIGURE 2.2 Aerial imagery with thermal data (hot spots) overlaid on it. This picture is provided with the permission of the New York State Office of Cyber Security and Critical Infrastructure Coordination.

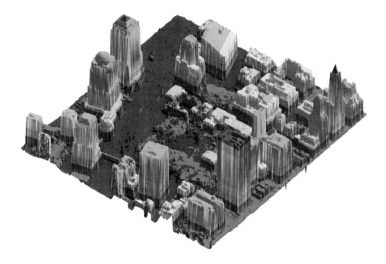

FIGURE 2.3 Aerial imagery depicting detailed elevation data obtained through the use of LIDAR (light detection and ranging) technology. This picture is provided with the permission of the New York State Office of Cyber Security and Critical Infrastructure Coordination.

and at least two detailed analyses have been published (Galloway, 2003; Thomas et al., 2003).[1] Several of these lessons relate to the use of geospatial data and tools, including

- The need for emergency management personnel to be aware of the value of geospatial information;
- The need for geographic information system (GIS) professionals who respond to emergencies to understand the needs of emergency responders and to have trained to meet those needs;
- The value of providing close geospatial support to first responders, sometimes referred to as a mobile or away team, which can be of assistance in providing the latest information, laying out search grids, analyzing data found, and transmitting event data back to the emergency operations center documenting the latest conditions at the incident site;
- The need to reduce the lag time in processing remote-sensing imagery, to avoid reducing its value for search and rescue and damage assessments;
- The need to have backup data for emergency management operations securely stored in widely distributed geographic locations;
- The need to have a preestablished list of geospatial professionals to provide support when needed during a catastrophe;
- The need to avoid duplication of effort and the confusion that results when the same products are generated by multiple groups; and
- The problems of relying on connectivity to the Internet in the immediate aftermath of a disaster, when it can be more efficient to distribute data by hand using physical storage media.

2.2 A HYPOTHETICAL CATEGORY 3 HURRICANE MAKING LANDFALL ON LONG ISLAND, NEW YORK

2.2.1 Background

It is late September, and a Category 4-5 hurricane has formed in the warmer-than-normal Atlantic Ocean and begins moving northward at a significant distance off the eastern seaboard of the United States. The entire East Coast of the United States from Jacksonville, Florida, to Montauk, New York, experiences high winds and surf, but no significant damage. The National Hurricane Center's forecasting models anticipate that it will weaken and turn easterly as it heads northward.

---

[1]See also *http://www.geoplace.com/uploads/featurearticle/0509ems.asp*.

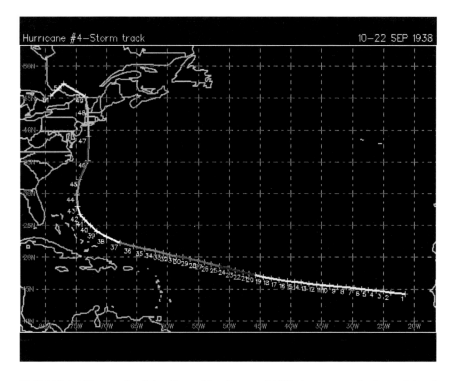

FIGURE 2.4 The track of the Great Hurricane of 1938, known as the Long Island Express, as it moved across the Atlantic Ocean and up the eastern seaboard. SOURCE: Figure created by Daniel Vietor, Unisys Corporation. Data from Atlantic Hurricane Database Re-analysis Project, National Oceanic and Atmospheric Administration.

Approximately 200 miles west of Bermuda, as predicted, the hurricane turns eastward. Its strength gradually drops to a Category 3 storm. Then, without warning, it turns northward again, intensifies, and begins to follow a track eerily similar to that of the great hurricane of 1938 known as the "Long Island Express" (Mandia, 2005) (Figure 2.4). Preparedness activities accelerate as the latest estimates predict landfall within the next 24 to 36 hours somewhere between New Jersey and Massachusetts. The storm accelerates as it continues northward, and landfall is now predicted in just over 12 hours. As the storm moves over cooler waters it again decreases in intensity to a Category 3 hurricane and makes landfall near Bellport, Long Island.

The hurricane quickly crosses over Long Island and makes a second landfall west of New Haven, Connecticut. Most of the south shore of Long

Island, the northern shore of Great South Bay, lower Manhattan, coastal Connecticut, and portions of Jersey City and Bayonne are covered by 7-10 feet of water. More than 8 feet of water accumulates at the entrance to the Lincoln Tunnel and floods the subway and rail tunnels. The region's three major airports, La Guardia, Kennedy, and Newark Liberty, are under water. Manhattan is virtually isolated, with bridges damaged by the high winds, and tunnels and bridge access routes flooded. Wind damage extends inland for hundreds of miles.

2.2.2 Pre-event Geospatial Preparedness

As the hurricane approaches, residents secure their homes, mandatory local evacuations ensue, and shelters are opened and staffed by Red Cross personnel. Natural disaster response plans (state, regional, county, and municipal) include geospatial data such as

- Locations of population and infrastructure, detailed maps providing estimates of the number of residents in each small area;
- Locations of critical infrastructure potentially impacted by the impending storm;
- Shelter locations, capacity, and service areas;
- Track of the hurricane overlaid on maps of populations at risk;
- Likely storm-surge inundation zones delineated using SLOSH (Sea, Lake, and Overland Surges from Hurricanes) models (storm surge), distributed to coastal counties and New York City (Figure 2.5);
- Road networks and capacity, egress routes, and traffic control points for the main state and county roads;
- Critical infrastructure data (pipelines, power generation) are available on a secure, on-line site for state officials, but not widely available to county or local governments; and
- The locations of special-needs residents, such as those in nursing, adult care, correction, mental health, and youth facilities that were identified in the hurricane plans. Hospital capacity data are also available, but are not current.

2.2.3 Likely Post-event Response

Once the hurricane has made landfall, telephone and network communications fail and most communities are unable to communicate with each other and with state or federal officials. Cell phones and radios do not function. Responders are unaware of evolving and new threats across the area, such as major flooding. They only know what they confront, but have no way to report those conditions. They begin to work on solving

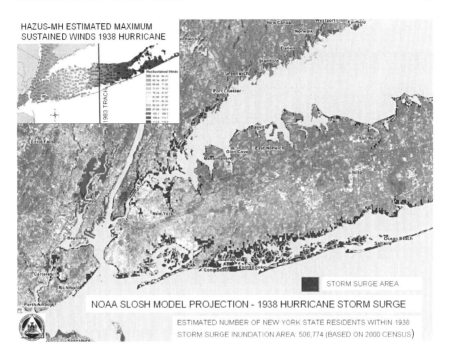

FIGURE 2.5 This graphic is intended to provide an example of the results of a SLOSH model. SOURCE: New York State Emergency Management Office GIS. Surge heights modeled by the National Oceanic and Atmospheric Administration.

problems where they are, but do not know how to locate additional resources or where to take the victims they rescue. By day two, responders have limited, intermittent radio communications, but there is so much radio traffic that they cannot be sure that the dispatch center hears what they report since they are never answered.

Because of the winds and the flooding, many streets are impassable while others no longer have valid signage or operational traffic control. Emergency responders have an urgent need for maps to organize search and rescue and initial response. Responders trying to rescue victims cannot get information about water depths, locations of high ground, and locations of dry streets so they can figure out where to launch boats and viable routes to and from launch sites. Responders feel disoriented even in areas with which they are very familiar because street signs and landmarks have been obliterated. After the fact, they realize that much better routes had been open and available, but they had no way of knowing this because they had little overhead intelligence other than from helicopters.

They thus have trouble getting aerial views of the disaster area so they can locate victims, prioritize rescue efforts, and direct ambulances. If they could see the current condition of the road system, they could deploy more efficiently and accurately, rather than by trial and error. They would also be able to identify open hospitals and other useful sites (e.g., sites where food and water are available). In short, they simply do not have a good clear picture of what things look like unless they go by boat or high-water truck. They do not have a web-based instant satellite or other real-time tool to use, even though their computers at the local EOC are up and could have been used.

As soon as they are able to fly, available state and local police helicopters are sent up with digital cameras to try to capture the damage. Reports of the extent of the damage are fairly good in New York City, but vary widely across Long Island. Requests are made by New York City and the counties on Long Island for immediate digital imagery to analyze the extent of the damage. Satellite imagery is not yet available because of cloud cover and because satellite overpasses are infrequent. Instead, requests for airborne imagery are made to New York State, to local imagery contractors, and to the federal government, leading to multiple flights over the area. The first digital imagery is provided five days later, although national news organizations showed visual images of the devastation from their news choppers much earlier. Given the time delay, local governments continue to use simple, digital cameras on the ground and in helicopters to record events as the recovery begins.

Some of the parties have previously shared copies of their framework data with New York State and have agreed which data will be used for analysis during the emergency. In the event that certain network or infrastructure restrictions limit data transfer, a secure site has been established at the U.S. Geological Survey (USGS) Center for Earth Resources Observation and Science (EROS) in Sioux Falls, South Dakota, to receive, store securely, and permit the sharing of incident data with authorized parties.

A request is made from Suffolk County on Long Island for additional GIS support. The state sends its "away" team to assist the county. This team is composed of skilled GIS professionals and brings with it a server filled with hundreds of state and local data sets, laptops, and a plotter. It is flown to the county by helicopter. It quickly sets up and begins producing maps and responding to inquiries for geospatial information such as coordinates for locations where rescues are still required. Other support is provided as needed from the state GIS team in the state emergency operations center. The teams fill many gaps for the county; however, their lack of connectivity with county networks results in delays in getting the information to the appropriate responders on the ground. Similar experiences are common in other affected counties and neighboring states.

Shortly after the first reports of the disaster are disseminated by the media, offers of assistance begin to flood in. Vendors of geospatial data and tools offer assistance of various kinds, some of it potentially helpful. However, no mechanisms exist for sorting through the offers, or for making use of the proffered data and tools in the chaotic situation that initially prevails. The Federal Emergency Management Agency (FEMA) deploys to this incident and sets up a Joint Field Office (JFO) with a GIS unit in the plans section. However this process takes several days and comes too late to be very helpful in the critical early stages of response.

2.2.4 What Could Have Been Done Better?

Few first responders have computer systems that allow them to share data directly with each other and the emergency managers in their jurisdictions. In some cases, emergency management officials do not even use the same geospatial data and tools as other agencies in their own jurisdiction, even in jurisdictions where enterprise-wide standards have been adopted. Such problems of interoperability may create barriers to communication between fire and police departments, emergency managers, and others and may result in many lost opportunities to take advantage of what responders on the scene are observing. Despite the promise of technology, therefore, much documentation of damage must still be done using paper forms.

Many of the geospatial data have been placed into various on-line servers by the respective agencies, to facilitate real-time access and updating by their own staffs as the emergency unfolds. All counties, New York City, and the states involved have GIS software, but it is not clear whether their data and systems can be shared or with whom. Overall, then, the picture is one of a patchwork of overlapping resources, developed in response to a variety of scenarios, some of which may involve sharing and integration of data between agencies and across jurisdictions. During an event of this magnitude, affecting such a large area and so many governments and jurisdictions, this patchwork of arrangements and geospatial preparedness would be wholly inadequate. While data and tools might be interoperable within one jurisdiction, the need for agencies to collaborate in unprecedented ways across jurisdictional boundaries will clearly pose problems, even if communication systems remain operable.

Critical gaps become evident in the geospatial data available for response, echoing the experience of Hurricane Katrina. For example, knowledge of the number and locations of households without private automobiles is a critical geospatial data need for evacuation planning. The availability, capacity, and locations of public transportation such as school buses, which could assist in evacuation, should be known. Planners

FIGURE 2.6 Web site reflecting the issues of pet evacuation during Hurricane Katrina. SOURCE: Animal Rescue New Orleans, *http://www.animalrescue neworleans.com.* Used with permission.

should anticipate the number and geographic distribution of household pets that would need sheltering. More importantly, the presence of pets in a household is a key determinant of a resident's willingness to comply with evacuation orders (see Figure 2.6), and geospatial data on the distribution of pets could therefore be very helpful in emergency preparedness. The notification and education of the public for evacuation and sheltering could be more effective if geospatial information such as maps were available in the form of pamphlets, on billboards, or in the local phone book.

Rapid procurement of digital imagery is critical during response. The area must be flown over and imaged almost immediately after the storm passes in order to capture the extent of the damage—such needs are predictable whatever the nature and extent of the event and should therefore be planned as part of geospatial preparedness. Furthermore, the digital data and images must be processed quickly to be of any use to emergency managers—certainly within 24 hours. This means that states and counties must have preexisting negotiated contracts with vendors or federal agencies that can be activated quickly to gather the field data.

While access to data *within* agencies was generally advanced, making extensive use of on-line servers, very little data sharing took place *between* agencies in advance of the disaster, and few of the problems associated with data sharing had been worked out. As a result, sharing had to be achieved ad hoc in the aftermath, when speed is critical, and problems had to be solved on the fly, leading inevitably to delays. Such delays could be minimized if sharing arrangements were worked out as part of geospatial preparedness. The expansion of existing data layers and incorporation of local data into an already robust, sharable database could help improve communications between the federal, state, and local emergency response communities.

The geospatial tools that would permit rapid analysis of data must be available to all agencies in all jurisdictions. For example, the timing of evacuations is critical and must be done in sequence in order to prevent massive traffic jams; geospatial tools exist to plan and manage such evacuations by identifying routes, simulating driver behavior, and planning traffic controls, but such tools are not widely available within the appropriate agencies. A secure on-line spatial decision support system designed for use by people with minimal GIS skills would facilitate emergency response decision making and provide a useful tool for local responders.[2] Tools that automatically provide Global Positioning System (GPS) coordinates to users would allow damage reports to be georeferenced quickly and easily, avoiding the need to wait until a trained GIS professional is available to perform the task.

The use of geospatial data and tools should have been fully integrated into emergency response across this region. Emergency responders need to be trained in the use of these data and tools, and steps have to be taken to incorporate them into the decision-making process. In addition, geospatial professionals who will be used to respond to such an emergency have to be identified, and around-the-clock contact information should be obtained for them as part of geospatial preparedness. It is also helpful to identify in advance the availability of volunteer geospatial expertise. Such expertise exists in many universities and in national and international volunteer relief organizations.[3] For example, the Computer Aided Design and GIS Research Laboratory at Louisiana State University was able to provide essential geospatial services in the aftermath of Hurricane Katrina.[4]

---

[2]During Hurricane Katrina, the State of Louisiana and New Orleans police shared 911 data on the location of victims through the simple-to-use web-based Homeland Security Information Network (HSIN).
[3]See, for example, *http://www.giscorps.org/*, *http://www.mapaction.org*.
[4]*http://katrina.lsu.edu*.

Land-parcel data, one of the framework themes, are essential in managing disasters and in assessing damage, along with building footprints and the locations of infrastructure (power, telecommunications, water, sewage, and steam-heating networks). GPS-enabled handheld devices for damage assessment will speed up the process of response and recovery as communities seek federal disaster aid. These computers should be programmed with "drop-down" menus to make it easy for staff to record data on storm damage, et cetera. Ideally, this equipment would have secure, wireless capability so that data can be forwarded electronically from the site to a location where they can be integrated with other data, to provide decision makers with a clear picture of the type and extent of impact in a given area. Doing this easily and quickly requires not only good planning, but also well-thought-out and frequent training.

Obtaining good data on the storm impact is important, but equally important is the distribution of these data to other response partners. If the Internet is accessible, this can be done most quickly through a secure, on-line application that provides web-based services, but if it is not, then other more traditional means of physical dissemination such as hand delivery will have to be used. Making the data available through a node of the National Spatial Data Infrastructure (NSDI) would allow them to be retrieved easily by authorized personnel across the country and would ensure compliance with standards. Smaller amounts of data might be e-mailed to a preestablished list of responders in neighboring counties and municipalities, state government, and the federal government.

Because of the large number of evacuees, the distribution of supplies and equipment to impacted areas must be well coordinated. The need for food, water, and so forth, for shelters housing large numbers of evacuees can be known in advance based on the population programmed for each shelter. However, the length of time that evacuees are required to stay in a shelter might vary depending on the damage to their home neighborhoods. Inventories can be managed and replenished if tools such as barcoding devices can be used to provide data to a central database, where they can be combined with preestablished geospatial data to support faster and more effective decision making. Finally, it is critical that all communities develop and maintain geospatial data on special-needs populations such as the elderly, infirm, or those without cars. These residents will require special planning and additional resources to move them out of harm's way.

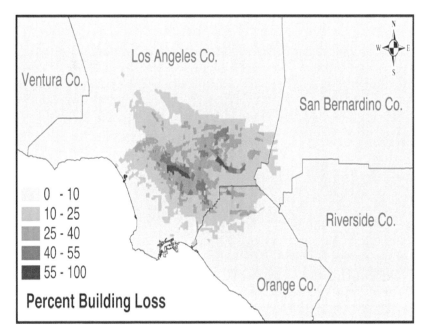

FIGURE 2.7 Damage based on the Puente Hills earthquake scenario. SOURCE: Southern California Earthquake Center (2005). Available on-line at *http://www.scec.org/core/public/image.php/13085-2162* [accessed on October 6, 2006]. Reprinted with permission of Earthquake Engineering Research Institute.

2.3 A HYPOTHETICAL SOUTHERN CALIFORNIA EARTHQUAKE

2.3.1 Background

It is another beautiful April Wednesday in Southern California with temperatures in the 70s. Roughly at midday, and without any warning, a rupture occurs on the Puente Hills fault, causing a magnitude 7.2 earthquake. Nearly 9.9 million people live in the Los Angeles Basin, and it is safe to assume that most of them feel the earthquake. Given the time of day, location, and magnitude, the early estimate of the number of fatalities is 7,600, with casualties running between 56,000 and 268,000.[5] The ground motion and shaking destroy more than 40 percent of the commercial and industrial facilities in the immediate area, as well as critical facilities such as hospitals and schools (Figure 2.7). Power and transmission

---

[5]The Southern California Earthquake Center projects losses of up to $250 billion for earthquakes on the Puente Hills fault under Los Angeles; see *http://www.scec.org/research/050525puentehills.htm* and Field et al. (2005).

lines are destroyed, and the basic transportation infrastructure is seriously compromised. Residential structures are also affected, resulting in upwards of 735,000 displaced households. Ruptured gas lines create the potential for fires, which break out sporadically in the affected area. The water supply is disrupted, with significant damage to the aqueducts bringing water from Northern California, the Owens Valley, and the Colorado River.

Short-term public sheltering is required for 211,000 people who are unable to move out of the area or to find alternative accommodation. Communications are difficult, given the 182 different languages spoken in the region; coupled with the loss of the major communications infrastructure, this poses significant rescue and relief challenges. Aftershocks continue for the next few days and keep the city on edge.

2.3.2 Pre-event Geospatial Preparedness

As part of their disaster preparedness, both the City of Los Angeles and the much larger County of Los Angeles have geospatial data and tools that are frequently used and constantly maintained, and emergency management personnel are regularly trained in their use. Examples of geospatial data helpful in preparedness for this event include the following:

- Locations of transportation routes, including road capacity, traffic flows based on time of day, and workarounds (alternate routing should there be an accident or some other traffic delay) are available in real time. Aerial reconnaissance from traffic helicopters operated by both public safety agencies and the major news media provide situational updates as required.
- Demographic data exist at fine levels of geographic detail, stored in databases and providing information on numbers of people and numbers of households, age characteristics, race and ethnicity, language spoken at home, number of persons with disabilities, number of households without cars, and number of persons living in poverty. These data are available at the block level for the city and county.
- Data on the locations of potable water lines, wastewater treatment facilities, sewer lines, natural gas lines, oil storage facilities, electric lines, communications lines, and cell phone towers (often termed *lifeline* data) have been culled from both private- and public-sector sources. In addition, geospatial data on critical infrastructure, including bridges, overpasses, dams and reservoirs, aqueducts, day care facilities, hospitals, schools, nursing homes, and urgent care centers, have been gathered and regularly maintained. Capacity data for hospitals are available, but there

THINKING ABOUT WORST CASES

is no up-to-date occupancy data. The same is true for schools, nursing homes, and day care facilities.

- Communication systems capable of reporting the user's location, accessing and displaying maps, and performing other simple functions on geospatial data are available on a limited basis for selected city and county emergency managers, but are not widely distributed to local first responders.
- Building inventory data is integrated with cadastral data for the entire county. Although residential areas are clearly identifiable, there is no consistent determination of the building's commercial or industrial usage.
- The locations of hazardous materials treatment, storage, and disposal facilities are available, but the exact quantity and type of materials at each location are not known. The city has a primary EOC that has all of the geospatial data on-site, backed up at three additional locations scattered throughout the region, but there is no backup facility outside the region.

2.3.3 Likely Post-event Geospatial Response

Most of the buildings near the epicenter have been completely destroyed, and the suspected area of damage extends outward for more than 50 miles (see Figures 2.8 and 2.9). The EOC sustained partial damage but

FIGURE 2.8 Apartment building collapse, Northridge, California, 1994. Photo credit, Susan Cutter.

FIGURE 2.9 Parking garage collapse, Northridge, California, 1994. Photo credit, Susan Cutter.

is functional in the short term. Sporadic reports begin coming into the EOC of catastrophic building failures within the county, especially among the older commercial and industrial buildings east of downtown Los Angeles. Power outages are widespread, and communications are spotty. Cell phone service is interrupted and, when available, is overloaded. Ruptured gas mains continue to pose fire hazards, and fires are burning out of control in some portions of the affected areas. Mass casualties are reported, and many of the major surface roads have suffered extensive damage.

Almost immediately, city and county officials begin airborne surveillance by helicopter. First responders begin fanning out to assist in search and rescue in their immediate vicinities, but there is little coordination. Search grids are laid out, but this has to be done by hand using paper maps because no power is available to run computers, printers, or copiers. Communications between first responders are limited, and mutual assistance from surrounding communities is not forthcoming due to the geographic extent of the damage.

Emergency power generation quickly enables some computer systems to become operational, providing access to geospatial data and tools. The region has an extensive network of trained geospatial professionals, especially among its HAZUS working group.[6] Calls go out to mobilize this

---

[6]HAZUS is FEMA's geospatial tool for estimating potential losses from disasters; see *http://www.fema.gov/hazus/*.

resource to help in the response. Since many responders cannot get to the EOC, secondary sites are established at major research universities where power is available, allowing some geospatial data and tools to be shared over the Internet.

2.3.4 What Could Have Been Done Better?

The mass casualties and injuries turn out to have been underestimated since the population estimates were based on nighttime residence (derived from census data), not on where people worked or attended school, and casualty rates in places of work and in schools turned out to be generally higher than in residential buildings. Having access to daytime as well as nighttime population data would be very helpful in responding to daytime disasters such as this. While HAZUS can generate daytime and nighttime impact values so that an archive of scenarios can be built, immediate access to daytime population estimates is just as important as access to census block data in case the scenarios do not cover actual conditions. Similarly, better data on individual buildings and the total numbers of people at work or visiting on any given workday would improve urban search-and-rescue efforts.

Ample geospatial resources were available elsewhere in Southern California, notably from universities and the private sector, which assisted greatly in the disaster response. However, these resources were not as effective as they might have been because of the lack of geospatial data backup at locations outside the affected area, and many of the data sets had to be reconstructed from other sources. The trained volunteer labor, however, proved useful.

Mobile, handheld devices for search and rescue and initial damage assessments by first responders would enhance the disaster response, especially if these could be enabled with simple geospatial data and tools, including GPS. Since many of the most damaged structures were commercial and industrial buildings whose footprints could extend to entire blocks, more precise locational information on inhabitants would be useful for search and rescue. For example, it would be useful to know the number of occupants of each room on each floor. Also important is knowledge of the exact location of potentially hazardous materials within these buildings, if any, so that fire and hazardous material (hazmat) teams can anticipate and deal with them effectively.

Remote-sensing imagery to gauge the magnitude and extent of damage, monitor particulate plumes from fires, and search for potential leakages in the numerous dams in the region could enhance response. Heroic efforts to acquire, process, and distribute imagery have been made by satellite operators, emergency managers, and geospatial professionals during recent disasters such as Hurricane Katrina or the events of September 11, 2001. Yet despite its potential, the immediate use of remote sensing remains difficult and limited, particularly due to a lack of equipment, training, and rapid access to suitable imagery, and this continues to thwart effective emergency response. To date, vendors and governments appear to have been unable to secure appropriate contracts for image acquisition and processing in anticipation of emergencies, leading to unacceptable delays in rapid damage assessment when events occur. These factors will continue to reduce and delay the use of remote sensing as a source of geospatial data during the response phase of disasters.

2.4 SUMMARY

All three of these cases, whether real or hypothetical, afford an opportunity to illustrate the value of geospatial data and tools in emergency management, to give examples of the current status of geospatial preparedness, and to ask what could be done better. Although these cases focus primarily on the initial response phase, geospatial data and tools also play vital roles in short- and long-term recovery. Both discussions of hypothetical scenarios included lists of geospatial data types that proved useful, and it is not surprising perhaps that the lists are remarkably similar—the geospatial data needed to respond to large-scale disasters will always include standard framework and foundation data sets, as well as extensive event-related data, although their availability and quality will vary markedly in practice. While the hypothetical scenarios were chosen to illustrate the differences between an event for which there was warning (the hurricane) and one for which there was none (the earthquake), in practice both scenarios proceeded in similar fashion. Geospatial data and tools played a significant role in the warning process, but problems such as lack of data interoperability, lack of training, and lack of effective communication rapidly reduced any benefits that advanced warning might have produced.

Virtually every aspect of emergency management requires knowledge of *where*: the location of the event itself, the magnitude of its impact on every part of the surrounding area, the locations of the assets that will be needed in the immediate response, the locations of infrastructure and evacuation routes—the list is endless. Effective emergency management also requires access to tools that can process such data to generate products of use to responders: maps of the area showing the search grid, images showing damage, lists of locations flagged for inspection, or predictions of the spread of wildfires or smoke plumes. The response phase of emergency management requires *speed*, in the form of fast access to up-to-the-minute data, rapid generation of key products, and rapid delivery to the personnel who need the products most. Accurate information delivered quickly can save lives, reduce damage, and reduce the costs associated with emergency response. Yet speed can be achieved only if responders are adequately prepared, through training, planning, and the coordinated development of procedures.

To assess the current status of geospatial readiness, one needs to ask the following questions. Are there significant gaps in the integration of geospatial data and tools into emergency management, and is the rapid delivery of geospatial information to key responders and decision makers during an emergency sufficiently straightforward? Do problems arise when disasters span many jurisdictions, because of lack of interoperability of geospatial data and tools and lack of foresight and experience in working together? Would better integration of geospatial information into the emergency management workflow improve decision making at all levels? Is there a need for better training in the use of geospatial data and tools among emergency management officials, and are geospatial professionals sufficiently trained in emergency management processes and practices? Is the rapid delivery of geospatial information a critical issue, and can emergency management workflows and standard operating procedures be redesigned to take advantage of this information? Finally, would regularly scheduled simulation exercises help all parties to learn how to meet each others' needs? These questions are very closely related to the committee's charge and are investigated in detail in the following chapters.

As noted elsewhere (Bruzewicz, 2003; Cutter, 2003; ESRI, 2001; FGDC, 2001; Goodchild, 2003; Greene, 2002), there are substantial challenges and constraints in using geospatial data to prepare for the worst cases. Many of them have already appeared in the examples described above and in the literature. In Chapter 3 the nature of disaster management and the relevant national policy context are reviewed in detail. Then the challenges and gaps in our current ability to prepare and respond with appropriate geospatial data and tools are summarized, as gathered from the literature and from testimony offered to the committee. In Chapter 4 the implications of these challenges and gaps are assessed in the light of current technology and anticipated future developments.

3

Emergency Management Framework

Intervention to address disasters has evolved through time into a complex policy subsystem, and disaster policy is implemented through a set of functions known as emergency management and response. Modern approaches to emergency management and response involve multidimensional efforts to reduce our vulnerability to hazards; to diminish the impact of disasters; and to prepare for, respond to, and recover from those that occur. These responsibilities present formidable challenges for governments because of the extraordinary demands disaster events impose on the decision-making systems and service delivery infrastructure of the communities they affect. Moreover, by definition an event constitutes a disaster if it exceeds the capacity of the government or governments in whose jurisdiction it occurs. Dealing with disaster therefore requires outside resources. In the context of a federally structured government, when the capacities of government jurisdictions at lower levels are overwhelmed, higher levels are called upon to assist, by either supporting or supplanting the activities of the subordinate jurisdictions. Likewise, assets and capabilities in the corporate and nongovernmental sectors may be brought to bear. As a result, emergency management and response are intrinsically intergovernmental, cross-sector policy implementation challenges. Also, since disasters dramatically affect our physical, social, and economic geography, geospatial requirements and capabilities are embedded throughout this complex system. This chapter describes the key characteristics of disasters and the conventional phased approach to their management, with particular attention to geospatial needs and functions.

3.1 THE CONTEXT OF DISASTERS

The paramount goal of disaster management activities is to reduce, as much as possible, the degree to which a community's condition is worsened by a disaster relative to its pre-disaster condition. There are many actions undertaken by participants in disaster management that support this goal both pre-disaster (to forestall or reduce potential damage) and post-disaster (to recover from actual damage), and ideally these activities would reduce the potential effects of a disaster to the point of elimination. Yet the very nature of disasters makes this ideal unachievable. There are five major characteristics of disasters that make them hard to overcome (for a more detailed explanation, see Donahue and Joyce, 2001; Waugh, 2000):

1. *Disasters are large, rapid-onset incidents relative to the size and resources of an affected jurisdiction.* That is, they harm a high percentage of the jurisdiction's property or population, and damage occurs quickly relative to the jurisdiction's ability to avert or avoid it. They may also directly impact the resources and personnel available to respond. As a result, response to disasters evokes a profound sense of urgency, and coping with them drains a jurisdiction's human resources, equipment, supplies, and funds. If pre-incident data are available, geospatial analysis can provide important insight into the nature and extent of changes wrought by disasters.

2. *Disasters are uncertain with respect to both their occurrences and their outcomes.* This uncertainty arises because hazards that present a threat of disaster are hard to identify, the causal relationship between hazards and disaster events is poorly understood, and risks are hard to measure—that is, it is difficult to specify what kind of damage is possible, how much damage is possible, and how likely it is that a given type and severity of damage will occur. Geospatial models can help predict the locations, footprints, times, and durations of events, and the damage they may cause, so that jurisdictions can better prepare for them.

3. *Risks and benefits are difficult to assess and compare.* Disasters present emergency planners, emergency managers, and policy makers with countervailing pressures. On the one hand, it is important to minimize the exposure of populations and infrastructure to hazards; on the other, people want to build and live in scenic, but hazard-prone, areas and often oppose government regulation. Further, how should the various levels of government address the balance between providing relief to the victims of disasters and the need or desire to avoid encouraging risk-accepting behavior; also, to what extent should the costs of such behavior be shifted from those who engage in this behavior to the larger population? While

most agree that response assistance should be provided to those who have suffered from a disaster, questions arise as to whether insurance for those in risk-prone areas should be subsidized by the federal government and to what extent repeated damage should be compensated (for example, by paying for rebuilding the same house after a second or third flood). An important component of this issue is the accuracy of risk assessment. Geospatial data and tools are invaluable in making the necessary assessments of the geographic distribution of risk and in estimating the quality of each assessment.

4. *Disasters are dynamic events.* Disasters evolve as they progress, and they change in response to human actions and natural forces. This makes it imperative that response strategies be flexible and argues for the value of analysis in helping responders understand and adapt to the changing conditions they face. Managing these phenomena can thus be a highly technical endeavor requiring specialized expertise for both policy development and policy implementation. In particular, geospatial data and tools can help incident managers to visualize the event over time, track the activities of responders, and predict the outcomes of various courses of action.

5. *Disasters are relatively rare.* Most communities experience few, if any, disasters during the average time in office of a political official or the average time of residence of a citizen. Thus, many communities are unlikely to have recent experience with disasters, and governments may feel little imperative to build their disaster-management capacity, even if the hazards are real and the risks formidable (Waugh, 1988). More obvious and immediately pressing public service concerns readily displace disaster preparedness as a priority. Specialized capabilities, such as geospatial data and tools, are especially vulnerable to budget cuts and resource reallocation.

These inherent qualities of disasters leave governments in a quandary about what to do to manage them. More specifically, the magnitude, scope, uncertainty, dynamism, and infrequency of disasters give rise to some important questions:

- How can we increase the resilience of communities to disasters—for example, by adding levees, raising the elevation of the living floor in homes, or imposing zoning regulations?
- How can we reduce the impact of disaster events—for example, through more effective warning systems or better evacuation plans?
- How can we most effectively provide assistance to those who have been affected—through development of a common operating pic-

ture and common situational awareness shared by all emergency responders or through better search-and-rescue procedures?

Thus, we face both policy issues and practical challenges as we work to reduce the risk to which our populations are exposed and to protect people and infrastructure. Almost every emergency preparedness and response challenge has important geospatial aspects, and effective emergency management thus requires adroit use of geospatial data and tools.

To address these and other issues and challenges, the emergency services professions have specified a host of activities aimed at assuaging the losses that disasters inflict. The degree to which these activities have been identified, assigned to responsible parties, and coordinated has evolved over time into a broad framework first defined in a 1979 National Governors Association report on its study of emergency preparedness (National Governors Association, 1979). This approach, known as Comprehensive Emergency Management, specifies four phases of modern disaster management: preparedness, response, recovery, and mitigation. Each of these phases levies particular demands on emergency managers and responders, and each can be informed and improved by the application of geospatial data and tools. These phases follow one another in a continuous cycle, with a disaster event occurring between the preparedness and the response phases, as shown in Figure 3.1. For additional explanation of the emergency management process, see Waugh (2000) and Haddow and Bullock (2003).

3.1.1 Preparedness

Preparedness involves activities undertaken in the short term before disaster strikes that enhance the readiness of organizations and communities to respond effectively. Preparedness actions shorten the time required for the subsequent response phase and potentially speed recovery as well. During this phase, hazards can be identified and plans developed to address response and recovery requirements. Disaster plans are often developed by individual agencies, but one challenge of disasters is that they demand action from agencies and organizations that may not work closely together from day to day. Thus, plans are much more effective when developed collectively by all agencies that will be responding so that resources and responsibilities are coordinated in advance. Also during the preparedness phase, training and exercises may be conducted to help prepare responders for real events. These vary from conceptual discussions to more formalized *tabletop exercises* (TTXs), during which neither people nor equipment is moved, to *field exercises* (FXs), which simu-

EMERGENCY MANAGEMENT FRAMEWORK   51

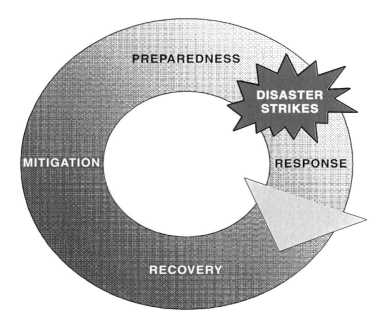

FIGURE 3.1 Emergency management cycle.

late real events. As with planning, training and exercises may be conducted by agencies in isolation, but they are more powerful when conducted jointly so that interfaces can be resolved. Perhaps the most important result of joint planning and exercising is the relationships developed between those who will be involved in response. In the best instances, these processes develop trust among those who will be called upon to work together during an event.

From the geospatial perspective, preparedness objectives include identifying data requirements, developing data sets, and sharing data across agencies. This includes activities as basic as developing framework data and foundation data on infrastructure, hazards and risks, location of assets that are of use for response and recovery (sand bags, generators, shelters, medical resources, heavy equipment, breathing apparatus, chemical spill response units, etc.), determining (if possible) common standards for data, making potentially difficult decisions about attributes, and compiling necessary metadata. Preparedness is greatly facilitated when all potential responding entities are working with the same data sets for the same features. Decisions also must be made as to whether data will be accessed from single sources or whether they will be hosted by some or all of the agencies involved in the response. Discussions about how geospatial support will be provided (each agency supporting its own geospatial work or some form of sharing of human resources) should occur. Applications, such as web servers and services and databases related to specific recovery and response activities, should be developed. Decisions should be made about how data are to be reported (times, units, method, format), which agencies will be preparing reports, and where the data and information are located and how and by whom they can be accessed. If imagery is to be used during the response, this is the time to consider user requirements for each mission, imagery that will meet these requirements, whether imagery may meet multiple requirements, what steps will facilitate the acquisition of this imagery, and how and to whom the imagery will be distributed after it has been acquired.

In the preparedness phase, geospatial tools can be used to display the distribution of hazards and risks as they exist now and risks as they may exist under different future development scenarios. This enables local and regional planners to work with emergency managers to plan for more sustainable futures through the avoidance or mitigation of higher-risk alternatives. For example, evacuation routes can be planned based upon demographics, capacity of existing roads, and traffic volume as a function of day and time.

Models of event scenarios can be used either in the development of single- or multiagency response plans or as part of exercises designed to test agency preparedness and the adequacy of those plans. The scenarios are essential in developing the master scenario events lists (MSELs) that enable exercise designers and controllers to test critical aspects of response plans and to develop additional modifications of the course of events during an exercise. Models also can be used prior to the actual impact of an event (pre-landfall for hurricanes or prior to flood crest) to estimate potential numbers of fatalities, injuries, and damage to infrastructure, so that responding agencies can initiate activities as soon as it is safe to move into the impacted area. Wind-speed models for hurricanes can be used to estimate the extent of expected damage to buildings. Energy-infrastructure damage models can be used to estimate the likely extent of damage to the distribution grid, and water- and ice-demand models can be used to estimate initial daily demand for these commodities.

3.1.2 Response

Response activities are undertaken immediately following a disaster to provide emergency assistance to victims. The response phase starts with the onset of the disaster and is devoted to reducing life-threatening conditions, providing life-sustaining aid, and stopping additional damage to property. During this phase, responders are engaged in a myriad of activities. As examples, search-and-rescue efforts are made to find individuals who may be trapped in buildings, under debris, or on roofs; basic commodities such as water and ice are distributed to affected populations; temporary power and shelters are established and provided; and fires and spills or leaks of hazardous materials are controlled. Although this phase is considered to begin when disaster strikes, not all disasters occur suddenly and without warning—sometimes onset is slower or anticipated, in which case response overlaps with the preceding preparedness phase and may include proactive steps such as warning and evacuation. Likewise, this phase has been defined historically as lasting 72 hours, but a clear end point for this period is difficult to define. It transitions into the recovery phase, and in reality response and recovery may overlap, especially during large, complex incidents.

Geospatial information and analysis are critical inputs to incident management and tactical decision making. Activities during this period include image acquisition, processing, analysis, distribution, and conversion to information products. Other geospatial data also must be collected, collated, summarized, and converted into maps, reports, and other information products. While sophisticated imagery and analysis are valuable to the response effort, the products most in demand are maps, including, for example, maps of the impact area and of the extent of damage; the locations of population in the impact area; the locations of assets to be used in the response, including inventories of critical supplies such as potable water and ice, temporary roofing material, medical supplies, and generators; maps of the area without power and of the timing of the return of power; and maps of road and bridge closures and downed power lines. Beyond this, products must also be useful and usable, which means that quality assurance and quality control (QA/QC) procedures and accurate metadata are essential. Attention must be given to reducing errors that arise when data are collected by different entities, or at different times, and then integrated into information products. Agreements need to be made regarding data reporting intervals and times, and data have to be time-stamped accurately. Finally, generation of data, information, and products is only part of the challenge—these must then be distributed to those who need them to do their jobs. Geospatial data are often voluminous, and this is especially true of imagery, which may amount to hundreds of megabytes or even gigabytes. Moving such volumes of data over networks that may have been partially disabled can be problematic, and Internet access to data repositories often fails. Firewalls and other security software installed on networks can also pose problems for the distribution of data and can significantly slow response. Agencies have often had to resort to physical distribution of CDs (compact discs) and other digital media during the response phase.

During the response phase immediately following an event, but prior to good information being available either from remote-sensing sources or from reporting on the ground, geospatial models can be used to provide damage estimates (e.g., immediately after an earthquake). Alternatively, real-time data from in situ monitoring can be used with geospatial models to determine conditions during an event, such as the use of real-time stream gauge data to issue flood warnings or the use of Doppler radar data, which results in the issuance of public warnings for severe thunderstorms and tornadic activity. While both imagery and verified reports from the impact area will eventually replace and refine the information provided by models, the latter may be the best source of information for several days after the onset of the disaster. Use of dynamic models can help guide and improve response; for example, the wildfire community makes extensive use of real-time and near-real-time geospatial modeling of wildfire behavior for logistical support. Display functions remain important at this time, showing the location of damage to specific infrastructure components (e.g., the transportation and energy infrastructure) as well as the severity of damage and other specific information (e.g., damage to roofs, temporary repairs, and energy grid restoration planned during the next 24 hours).

Accomplishing all of these tasks is admittedly a substantial challenge in the earliest stages of disaster response, when demands are urgent and requests are voluminous. Poor products can have serious negative ramifications for response and recovery operations, however. For geospatial professionals to perform well in this environment, they must be able to rely on good training, relevant exercise experience, and sound standard operating procedures.

3.1.3 Recovery

Recovery includes short- and long-term activities undertaken after a disaster that are designed to return the people and property in an affected community to at least their pre-disaster condition of well-being. In the immediate term, activities include the provision of temporary housing, temporary roofing, financial assistance, and initial restoration of services and infrastructure repair. Longer-term activities involve rebuilding and reconstruction of physical, economic, and social infrastructure and, ultimately, memorializing the losses from the event.

Geospatial activities during recovery include the use of geospatial information and analysis to help managers direct the recovery process, including the urban search-and-rescue grid and status, tracking the progress of repairs, provision of temporary water and ice, locating populations, identifying sites for temporary housing and services, and showing the operational status of hospitals and clinics.

An important task is capturing and archiving data collected as part of the disaster, along with copies or descriptions of the procedures that were used to turn those data into information and to distribute the information, and documentation of lessons learned from the disaster. These data can be used to inform mitigation planning and research about disaster processes. Too often, however, archiving is given short shrift and valuable data are lost.

3.1.4 Mitigation

Mitigation includes those activities undertaken in the long term after one disaster and before another strikes that are designed to prevent emergencies and to reduce the damage resulting from those that occur, including identifying and modifying hazards, assessing and reducing vulnerability to risks, and diffusing potential losses. In short, it is a set of sustained activities designed to reduce the impacts of future disasters. Mitigation involves implementing policy changes and new strategies. Some of these activities may be structural in nature, such as changing building codes (e.g., to require that residential buildings be able to resist sustained wind speeds of 150 miles per hour [mph] rather than 120 mph, to require fastening roofs to bearing walls). Mitigation measures also can be nonstructural. For example, zoning can be used to preclude development in areas that are subject to risk from a hazard.

Geospatial assets can inform mitigation planning in important ways, perhaps most importantly the opportunity to visualize and measure the effects of alternative mitigation plans. Simulation models (e.g., to model the inundation area that will result from various stream elevations with and without the presence of levees or to predict the propagation of hazardous materials in the atmosphere) can help planners make redevelopment decisions. Geospatial analysis can support benefit-cost analysis by comparing the cost of changes (such as new construction requirements) to estimates of the savings that result when a hazard is mitigated. Geospatial tools are of particular benefit due to their ability to permit the evaluation of multiple alternatives relatively rapidly.

3.1.5 Additional Comments

The cycle shown in Figure 3.1 is clearly simplified, since events can occur at any time and may overlap. Different organizations come into play in different phases, creating a complex web of interactions. Recovery and mitigation may not be complete before another event occurs, and the necessary funds to support them may not be fully available prior to the next event. Further, as the ability to organize multiagency efforts continues to improve, some of the actions that have traditionally been thought of as recovery activities are now beginning at essentially the same time as the response. In theory, preparedness should reduce the time from the initiation of response to the end of recovery. Mitigation should reduce the cost of future disasters of the same type at the same location, and lessons learned should be incorporated into planning and mitigation in other areas, thereby reducing impacts elsewhere.

A modification of this paradigm is used for acts of terrorism where awareness, detection, deterrence, and prevention are seen as the key elements in reducing or eliminating the impacts or even the occurrence of events. Specific emergency management activities may differ for those described above as they are influenced by the intelligence and security communities, but the sequence is analogous to that followed for natural disasters and has elements that parallel what is required for technological disasters. For these events, intelligence must be collected about risks posed by individuals and groups that may seek to harm people or critical infrastructure. In parallel to preparedness and mitigation, techniques are developed to deter or reduce the effectiveness of attacks so that the consequences are reduced. In ideal cases, populations and infrastructure are rendered invulnerable to attacks. Again, geospatial data and tools can be used to show conditions at particular points in time. It is possible to model the consequences of various forcing mechanisms (attacks rather than wind speed or flooding) on the existing infrastructure under a range of response and deterrence mechanisms.

3.2 RELEVANT ACTORS

3.2.1 Emergency Managers and Responders

The catastrophic nature of disasters means that all levels of government and all sectors of society share responsibility for dealing with them. In general, disasters are managed through a federal structure of responsibilities and resources, where discretion and authority for management reside with the affected jurisdictions, and where requests for resource support travel upward from those jurisdictions until enough are garnered to stabilize the incident. Table 3.1 identifies the major functions of each level of government during each phase of the disaster management process.

It is often said that "disasters begin and end at the local level." The effects of disasters are felt by people living in communities, and ultimately the efforts of emergency services professionals focus on restoring the health of communities. Disasters are fundamentally local in impact; thus, responsibility for the management of response resides with states and local governments. Local responders provide the first response in communities, focused on initial efforts to save lives and property. As jurisdictions are overwhelmed, neighboring jurisdictions may assist through the provision of mutual aid. Nongovernmental organizations (both private and nonprofit) also supplement response with a range of assistance from providing shelter and food to helping manage donations of money, goods, and services, to tracking and serving populations with special needs. Figure 3.2 describes the sequence of events for response.

For larger incidents, impacts can extend to regional or even national levels, as was the case with Hurricanes Katrina, Rita, and Wilma in 2005, and the Space Shuttle *Columbia* crash in 2003. If local jurisdictions find they cannot manage the demands of an incident, they turn to their state government for assistance. State emergency managers coordinate local communities, state agencies, assets controlled by the governor (such as the state national guard), and support from other states and the federal government. They assess damage and resource needs, and then obtain and allocate required resources. If the size and scope of the incident warrant, the governor may request a disaster declaration.

In the event of a request for a disaster declaration, or if the disaster is national in significance or scope, the President may decide to bring the resources of the federal government to bear. These resources are generally coordinated by the Department of Homeland Security (DHS) under the National Response Plan (NRP).[1] The NRP "is an all-hazards plan that provides the structure and mechanisms for national level policy and operational coordination for domestic incident management." It provides the framework for federal interaction with other levels of government and other sectors with respect to all phases of disaster management (preparedness, response, recovery, and mitigation) and describes federal capabilities, resources, agency roles, and responsibilities.

One critical function of emergency responders at all levels of government is incident management. Incident management refers to the collection of command-and-control activities exercised to prepare and execute plans and orders designed to respond to and recover from the effects of an emergency event. It is usually effected through a functionally oriented incident command system (ICS) that can be tailored to the type, scope, magnitude, complexity, and management needs of the incident and can operate at all levels of government. An ICS is employed to organize and unify multiple disciplines, jurisdictions, and responsibilities on-scene un-

---

[1] *http://www.dhs.gov/interweb/assetlibrary/NRP_FullText.pdf.*

TABLE 3.1 Key Disaster-Related Functions by Level of Government and Phase

Level	Mitigation	Preparedness
Federal	• Supports research of hazard causes • Develops means to modify the causes of or vulnerability to hazards • Reviews and approves state mitigation projects • Provides training and technical expertise • Directs flood control program • Directs hazard prediction and mapping initiatives • Provides hazard mitigation grants • Provides funds to individuals for small projects to prevent losses • Funds coastal land-use planning • Creates geospatial data model • Provides federal flood insurance • Invests in development of new technologies	• Provides training and professional development programs • Provides public education • Coordinates warning system • Formulates, implements, and evaluates emergency management policy • Conducts inspection and assessment programs • Reviews, coordinates, and conducts federal, state, and regional exercises • Assesses and coordinates disaster plans • Provides grants for disaster planning, equipment, and training • Operates the national operations center • Specifies required response capabilities • Facilitates information sharing • Coordinates incident response planning • Synthesized intelligence • Generates threat assessments • Inventories critical infrastructure • Stockpiles equipment and supplies
State	• Conducts hazard identification • Conducts land-use planning • Develops, adopts, and enforces land-use standards • Regulates growth • Solicits mitigation projects and establishes funding priorities • Establishes legal basis for local ordinances • Regulates construction • Provides aid to localities	• Conducts risk and exposure assessment • Monitors and surveys potential hazards • Creates resource inventory • Conducts disaster planning • Coordinates plans of localities, facilitates interagency policy coordination • Stockpiles equipment and supplies • Conducts capability assessment • Provides public education • Conducts training and exercises • Provides technical expertise to localities • Obtains grant funding to support preparedness activities

Response	Recovery
- Collects data about the disaster - Creates and disseminates common operating picture - Assesses damage - President may declare disaster or emergency - Implements the National Response Plan and activates Emergency Support Functions - Designates principal federal official - Establishes Joint Field Offices to coordinate support - Provides atmospheric modeling - Can mobilize the military - Validates and makes recommendations in response to threat assessments - Provides food, water, temporary power, and technical assistance	- Restores economic stability - Provides crisis counseling - Provides legal assistance - Provides technical assistance, debris removal, communications, and public transportation, if requested - Provides temporary housing assistance, individual and family grants, funds to repair facilities, and disaster unemployment assistance - Provide loans for repair of homes, businesses, farms - Provides tax relief
- Mobilizes National Guard - Provides food, water, clothing, and shelter - Conducts damage assessment - Disseminates public information - Restores essential infrastructure - Executes state emergency plan - May request FEMA to assess damage - May seek presidential declaration - Runs EOC - Coordinates resources across jurisdictions - Funds mutual aid to other states - Provides aid to localities - Assists with evacuation	- Conducts debris removal - Restores public services and facilities - Restores infrastructure - Restores economic stability - Renews economic development - Restores governmental self-sufficiency - Prepares hazard mitigation plan - May request federal agencies to perform short-term tasks - Administers federal assistance - Provides technical assistance to localities - Provides relief funds to localities

continued

TABLE 3.1 Continued

Level	Mitigation	Preparedness
Local	• Controls siting of structures to avoid disasters • Develops, adopts, and enforces building codes and land-use standards • Requires construction of disaster-resistant structures • Initiates retro-engineering activities to correct inappropriate building designs • Regulates growth • Undertakes hazard identification and control efforts	• Analyzes and monitors hazards • Identifies and assesses risks and exposure • Identifies and inventories resources • Conducts disaster planning • Develops interagency and interjurisdiction response systems • Stockpiles, pre-positions, and maintains emergency equipment and supplies • Measures and assesses response capability • Conducts training, exercises, testing • Provides early warning • Conducts pre-disaster evacuation • Provides public education information

NOTE: EOC = emergency operations center; FEMA = Federal Emergency Management Agency.
SOURCE: Adapted from Donahue and Joyce (2001).

der one functional organization. The ICS establishes lines of supervisory authority and formal reporting relationships, but allows for team-based leadership approaches. In particular, the ICS may include the adoption of a formal unified command, a multiagency governance structure that incorporates officials from agencies with jurisdictional or functional responsibility at the incident scene and allows them to provide management and direction jointly within a commonly conceived set of incident objectives and strategies. Regardless of whether the ICS is configured as a unitary or a unified command, the ICS organization develops around five major functions that are required for any incident whether it is large or small:

1. Command. The incident commander's (IC's) responsibility is overall management of the incident. On most incidents the command activity is carried out by a single IC. The IC determines incident objectives and strategy, sets immediate priorities, establishes an appropriate organization, authorizes an Incident Action Plan, coordinates activity for all com-

Response	Recovery
• Warns public • Provides emergency communications • Evacuates public • Conducts search and rescue • Manages hazardous materials • Stabilizes debris • Provides emergency food, water, shelter • Disseminates public information • Restores essential infrastructure • Provides fire suppression • Provides law enforcement • Provides triage and medical care • Implements curfews • Funds ongoing emergency activities • Provides mutual aid to other localities	• Removes debris • Restores public services and facilities • Restores individual emotional health • Restores economic stability • Restores governmental self-sufficiency • Restores individual self-sufficiency • Rebuilds and repairs capital stocks, homes, and businesses • Restores infrastructure • Coordinates with nonprofit agencies to support human welfare activities • Renews economic development efforts

mand and general staff, ensures safety, coordinates with key people and officials, authorizes release of information to the news media and the public, and performs other key duties.

2. Operations. Operations refers to the ways in which resources are applied in the field to meet emergency response objectives. In an ICS, the operations section is responsible for directing and supervising the execution of all tactical activities. Operations chiefs also coordinate activities with other entities, ensure safety, and request and release resources. Operations are often facilitated by an associated planning process.

3. Planning. This function involves the collection, evaluation, processing, and dissemination of resource and situational incident information. This information informs the Incident Action Plan, which specifies how all incident operations will proceed. Geospatial assets are typically incorporated as part of the plans section, often through the use of technical specialists who provide a particular level of expertise necessary to properly manage the incident.

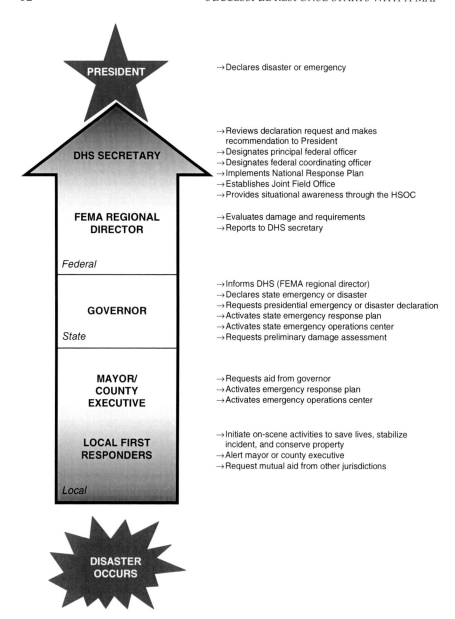

FIGURE 3.2 Emergency response and disaster declaration process. NOTE: DHS = Department of Homeland Security; FEMA = Federal Emergency Management Agency; HSOC = Homeland Security Operations Center.

4. Logistics. This function incorporates all incident support needs, including supplies, facilities, transportation, communications, food, and medical support. It is the logistics section's responsibility to establish the infrastructure required to meet the data management needs that arise as geospatial data and tools are brought to bear on an incident.

5. Finance and Administration. This function includes activities such as procurement, timekeeping, compensation, claims processing, and cost management. Should geospatial data and tools be required to support an incident, this section would be responsible for procuring them.

While the concept of incident command has been developed over more than three decades and is broadly employed, different disciplines and jurisdictions understand and implement ICS differently. Also, as the committee heard in accounts of incident after incident, the implementation of a coherent command structure for a large-scale disaster is a substantial challenge. Very often, multiple, overlapping, duplicative, and even conflicting command processes and structures emerge. This, in turn, makes coordination and application of geospatial resources difficult. The National Incident Management System (discussed below) attempts to address these tensions by incorporating longstanding ICS and unified command principles into a common incident management operating philosophy.

3.2.2 Public Sector Geospatial Support

Emergency services professionals, from those operating tactically at the front line to those working strategically at higher levels of government, are aided in their incident management responsibilities by a variety of geospatial experts using data and tools and working at various sites to support response and recovery. In the public sector, these geospatial experts reside in numerous federal agencies and national laboratories and in state and local governments. The role that each plays in emergency management activities varies according to the mission of the organization and is described in this section.

The federal agency whose mission is most closely involved in emergency management is the Department of Homeland Security. Since this agency was established in 2002, its development and operation of disaster management policy and functions are dynamic and still evolving. Nevertheless, from the geospatial policy perspective, DHS's Geospatial Management Office (GMO) was established by the Intelligence Reform and Terrorism Prevention Act of 2004,[2] section 8201 "Homeland Secu-

---

[2]*http://www.fas.org/irp/congress/2004_rpt/h108-796.html.*

rity Geospatial Information." The GMO is responsible for (1) coordinating the geospatial information needs and activities of the department; (2) implementing standards to facilitate the interoperability of geospatial information pertaining to homeland security among all users of such information within DHS, state and local governments, and the private sector; (3) coordinating with the Federal Geographic Data Committee and carrying out the responsibilities of DHS pursuant to Office of Management and Budget (OMB) Circular A-16 and Executive Order 12906; and (4) making recommendations to the secretary and the executive director of the Office for State and Local Government Coordination and Preparedness on awarding grants to fund the creation of geospatial data and execute information-sharing agreements regarding geospatial data with state, local, and tribal governments.

The GMO's major initiatives include publication of a draft Geospatial Data Model (May 2006),[3] developing geospatial guidance for DHS's grant program, developing a geospatial concept of operations for the National Response Plan, and publication of a national geospatial strategy to meet national geospatial preparedness needs. Of note is that by 2007, the office is supposed to be able to provide oversight of all geospatial IT (information technology) systems management, procurement, security, and interoperability issues at DHS.

The office has only a handful of staff and is currently funded at about $13 million. Given its broad mission and ambitious agenda, these resources seem inadequate. In fact, the DHS inspector general has noted that the GMO has "used a 'do no harm' approach—leaving legacy agencies within DHS . . . to manage as they deem appropriate," with the result that other DHS component managers are stalled, unsure when to coordinate with the GMO and when to act on their own (DHS, 2005, p. 22). This may be a symptom of the fact that the office is underfunded and understaffed. It is the sense of the committee that this office and its initiatives are relatively new and have not yet matured into a robust geospatial organization at DHS. It will be very important for geospatial capacity to be a strong component of DHS activities, both operationally (as part of the National Operations Center described below) and analytically (as part of efforts such as infrastructure protection).

Other DHS agencies also are involved in geospatial activities. Since July 2004, DHS has continuously operated a standing Homeland Security Operations Center (HSOC), designed to serve as a center for information sharing and domestic incident management and to help coordination be-

---

[3]http://www.fgdc.gov/fgdc-news/geo-data-model/.

tween federal, state, territorial, tribal, local, and private-sector entities. The White House views the HSOC as the primary federal-level command and control "hub for operational communications, information sharing and situational awareness for all information pertaining to a domestic terrorist or disaster incident." The HSOC collects and fuses information from a variety of sources and provides real-time situational awareness and monitoring nationwide. It also coordinates incidents and response activities and, in conjunction with the DHS Office of Information Analysis, issues advisories and bulletins concerning threats and specific protective measures. Information on domestic incident management is shared with emergency operations centers (EOCs) at all levels using the Homeland Security Information Network (HSIN). The HSOC reports receiving hundreds of calls and managing about 22 cases per day. To accomplish its missions, HSOC relies on **"watchstanders"** from 35 agencies, including the National Geospatial-Intelligence Agency, the National Oceanic Atmospheric Administration (NOAA), and the DHS Geospatial Management Office. The efficacy of the center has come under scrutiny since Hurricane Katrina. The 2006 report by Congress, *A Failure of Initiative: Final Report of the Select Bipartisan Committee to Investigate the Preparation for and Response to Hurricane Katrina*,[4] for example, notes that the center "failed to provide valuable situational information to the White House and key operational officials" during the storm. DHS is currently considering consolidating HSOC functions with those of other federal operations centers to form a National Operations Center.

In direct support of the HSOC and incident management, DHS relies on the ability to formulate a common operating picture (COP), driven by fused data sources and types and shared geospatial data and services, in order to facilitate situational awareness. Development of the COP is an important job of the GMO. In the future, DHS hopes to use its COP capabilities to support forecasting, modeling, and decision aids.

DHS also relies on its Interagency Modeling and Atmospheric Assessment Center (IMAAC) to serve as the "central provider" of atmospheric dispersion and hazard predictions for use by federal agencies and by state and local governments during incidents of national significance. The IMAAC is designed to assess atmospheric hazards relating to releases of chemical, biological, radiological, nuclear, and/or high-explosive materials. Following an atmospheric release event, IMAAC produces, coordinates, and disseminates consequence predictions in near real time to federal, state, and local responding agencies. In this way, responders will

---

[4]*http://www.house.gov/transportation/fullchearings/02-16-06/katrina_report.pdf.*

share an up-to-date common operating picture based on accurate, unambiguous hazard prediction models. DHS's vision is that the modeling products produced by IMAAC will be based on a combination of location-specific meteorological data, demographic data, and on-site observations garnered through the collaboration of various federal partners. The IMAAC concept has not yet been fully developed or implemented. In the interim, Lawrence Livermore National Laboratory's National Atmospheric Release Advisory Center is serving as a technical and functional example for IMAAC.

Geospatial professionals directly involved in emergency management work primarily in other capacities during normal times outside of disaster response and recovery, although this varies by agency and level of government. At the federal level, the Federal Emergency Management Agency (FEMA) has geospatial professionals who produce maps and analyses at the Mapping and Analysis Center (MAC) located at FEMA headquarters. This group generates products that track damage, demographic changes in damage areas, and other information of interest to policy makers. Staffing of this group is relatively small (approximately five people). In the field during an incident, FEMA supports its Joint Field Offices (JFOs) with a team of geospatial professionals as part of the plans section. From testimony, the committee learned that most of FEMA's geospatial personnel are not full-time permanent employees but reservists with geospatial skills who are called upon to staff JFOs during events. The number of reservists available has varied, and recent experience during the 2004 and 2005 hurricane seasons indicates that the number of trained professionals who can provide sustained support for large events is insufficient. Moreover, because reservists change from event to event, it is difficult for FEMA to build robust, stable geospatial teams that work together well and have a consistent understanding of their mission. Finally, it usually takes several days or even weeks for FEMA's geographic information system (GIS) section to be established at the JFO. Moreover, the GIS section does not deploy as a unit that is ready to work when it arrives on-site. Instead, a GIS team leader deploys and then assembles the personnel and equipment needed for the section. As a result, FEMA's ability to provide geospatial support can be significantly delayed. Delays in getting adequate equipment, data, and software on-site often further hinder FEMA's GIS support at the JFO. Meanwhile, state and local geospatial staffs have already begun work coordinating geospatial resources and developing products. It can be difficult for FEMA's GIS section to integrate its support with these ongoing efforts. In addition, some response problems end up being solved without the benefit of technology that FEMA might have brought to bear, had it been on-site and operational sooner. The committee was told that in 2006, FEMA would be testing a geospatial strike team approach by sending professionals to a disaster area to meet immediate mapping and analysis needs. The strike team would move out to be replaced by longer-term personnel as response and recovery activities proceed.

Other federal agencies and national laboratories also have geospatial resources, including personnel, data, and tools that can be brought to bear during disasters. The roles of the other federal agencies in emergency management and their use of geospatial data and tools are very diverse, depending on the mission of the agency. One important player is the U.S. Army Corps of Engineers, which has no geospatial professionals dedicated to emergency management, but has created a cadre of GIS professionals trained in the missions performed by the Corps under the National Response Plan and in support to FEMA at the JFOs. The National Oceanic and Atmospheric Administration is very involved in weather-related emergencies, providing a wealth of geospatial data and tools. The NOAA National Weather Service GIS Data web site[5] is but one example of the types of data that it makes available. The National Geospatial-Intelligence Agency (NGA) can be tasked by FEMA to provide imagery and geospatial analysis for disaster response, as it did during Hurricane Katrina. The U.S. Geological Survey (USGS) makes all of its geospatial data available through various means for use during incidents and has been asked by FEMA to help coordinate geospatial data efforts among federal agencies. A recent example of this is the partnership between USGS, NGA, and DHS to collect data for hurricane response, called GIS for the Gulf, which is being made accessible through the Geospatial One-Stop. The Environmental Protection Agency (EPA) performs various functions such as removing hazardous wastes from sites, risk assessments, and environmental sampling of drinking water, air, and sediments, and uses geospatial data and tools for supporting these functions. The U.S. Forest Service is very advanced in the use of geospatial data, tools, and models for wildfire response. National Aeronautics and Space Administration (NASA) remote-sensing data are used not only for response activities, but also for monitoring, modeling, and predicting natural phenomena. NASA works collaboratively with other agencies in developing applications and tools for disaster management. Although there are many more, these examples provide a feel for the broad range of geospatial activities related to emergency response carried out by the federal agencies.

At the local level, municipalities may have up to three dedicated geospatial professionals for emergency management activities, but most

---

[5]*http://www.weather.gov/gis/.*

small towns have none. Large cities often have their own GIS departments, and cities such as New York have GIS operations that are dedicated to emergency management. State emergency management agencies typically have up to seven geospatial professionals, but also often have their own GIS operations for other purposes such as environmental management or conservation and may draw upon this group of professionals during a disaster event.

Aside from the direct technical support that geospatial professionals provide, they are also in a position to play an important role in helping disaster managers and incident commanders think about how to define their geospatial requirements. In some cases, response professionals are not familiar with imagery, so they are unsure what is available that can help them, or they are unable to frame their questions effectively. In other cases, responders "have just enough information to be dangerous," thinking they need something that will not actually help to solve their problems. Also, sometimes responders ask for the wrong thing, which then fails, leading them to condemn geospatial technology as unworkable or inappropriate. To bring geospatial data and tools to bear effectively then, geospatial professionals must be conversant in how to define functional requirements and must be willing and able to engage their responder customers in a way that will help the customers to articulate their problems and needs, so that the GIS unit can respond to them.

3.2.3 Geospatial Support from Universities and the Private Sector

Another repository of relevant expertise is at universities and private-sector companies. Researchers and graduate students often support city, regional, and state planning departments, and they may become fully integrated into activities occurring during a disaster. Similarly, private-sector corporations often provide professionals to assist governments in their response. For example, following the events of September 11, 2001, assorted specialists and volunteers from universities relocated to Pier 92 to contribute to the city's geospatial response component, and various private-sector organizations proffered an array of technical capabilities and products to FEMA and the City of New York (see Section 2.1.1).

Private industry is also a large provider of geospatial information for emergency management, primarily in the area of aerial photography or other airborne remote sensing (light detection and ranging [LIDAR], thermal imaging, hyperspectral data) and satellite imagery. These data are vital for assessing impacts and damage immediately following an event and for supporting response and recovery activities. They also can be used for risk and vulnerability assessments in mitigation and preparedness phases. Section 4.6 describes the use of these types of data in more detail.

Some of the data necessary for emergency response, such as critical infrastructure data, are developed and maintained by the private sector.[6] Estimates from the Department of Homeland Security's Protected Critical Infrastructure Information (PCII) Program are that the private sector owns and operates 85 percent of the nation's critical infrastructure.[7] However, use of private-sector utility data is often very restricted. Issues with the use of these types of data are discussed in Chapter 4.

3.2.4 International Activities

The committee also queried members of the international emergency response community about geospatial needs. They largely echoed the needs expressed by responders in the United States and, in addition, cited key gaps in the global coverage of framework data and problems in preparing for and responding to emergencies that span national boundaries. Also, metadata at multiple levels are lacking, and as a result, there is poor harmonization among users. Attempts have been and are being made to address this problem.

There are numerous international activities in the area of geospatial data for emergency response, and much could be learned from them. Although it was beyond the scope of this study to analyze these activities in detail, a few are mentioned here as examples. The European Union Research and Development Programme has had programs running for many years on the specific role of geospatial information in disaster management, particularly in the Sixth Framework Programme.[8] The Emergency Planning College in the United Kingdom has recently published *A Guide to GIS Applications in Integrated Emergency Management* (MacFarlane, 2005). International efforts to support the development of a Global Disaster Information Network[9] (GDIN) were initiated in 1998. Subsequent meetings have been held to bolster this effort, and GDIN now exists as a voluntary self-sustaining nonprofit association. Moreover the Indian Ocean tsunami of 2004 created significant international interest in a warning network for such disasters. An excellent summary of lessons learned from the Indian Ocean tsunami regarding the deployment of geospatial data and tools in sudden-onset emergencies can be found in Kelmelis et al. (2006).

---

[6]*http://www.dhs.gov/dhspublic/display?theme=92&content=3760/.*
[7]See DHS PCII Program overview at *http://www.dhs.gov/dhspublic/interapp/editorial/editorial_0465.xml.*
[8]*http://cordis.europa.eu/fp6/dc/index.cfm?fuseaction=UserSite.FP6HomePage.*
[9]*http://www.gdin.org.*

The Civil Military Emergency Preparedness Program is an ongoing effort funded by the Department of Defense that is focused on former Warsaw Pact countries. This program has developed an on-line geospatial data and map server available through the Partnership for Peace (PfP) Information Management System[10] (PIMS). It presently includes an interface that provides multiscale access to data for these countries at a variety of levels of geographic detail from 1:2,000,000 to 1:10,000 that have been contributed by some of the participating countries. Low- and high-resolution imagery is also available. Other important activities include the North Atlantic Treaty Organization's (NATO's) Euro Atlantic Disaster Response Coordination Center[11] (EADRCC) located at NATO Headquarters in Brussels, and the United Nations' nongeospatial database of response resources that has been provided by each participating nation.

3.3 FEDERAL POLICY RELEVANT TO GEOSPATIAL REQUIREMENTS

As described above, demands imposed by disasters have prompted the evolution of emergency management into a formal set of activities assigned to responsible parties and coordinated across governments. Practices and policies have evolved over decades, and organizations and agencies, such as the Red Cross and Civil Defense, and emergency services, such as police and fire departments, have evolved a complex system of practices and procedures. However, the events of September 11, 2001, further crystallized these responsibilities. The Bush administration immediately established an Office of Homeland Security, and slightly more than a year later, the Department of Homeland Security was created. The DHS mission is stated as follows: "We will lead the unified national effort to secure America. We will prevent and deter terrorist attacks and protect against and respond to threats and hazards to the nation. We will ensure safe and secure borders, welcome lawful immigrants and visitors, and promote the free-flow of commerce." To meet this mission, DHS has implemented a National Incident Management System[12] (NIMS) and an updated NRP.[13] These and other recent federal-level policy documents explicitly recognize the requirement to make geospatial data and tools available to support incident management. This section will identify the major policies and plans that currently guide

---

[10]http://www.pims.org.
[11]http://www.nato.int/eadrcc/home.htm.
[12]http://www.fema.gov/nims/.
[13]http://www.dhs.gov/interweb/assetlibrary/NRP_FullText.pdf.

incident management and emergency response and their requirements related to geospatial data and tools.

3.3.1 National Strategy for Homeland Security[14]

This strategy, issued by President Bush in July 2002, sought to articulate the administration's vision for homeland security policy and to provide direction and guidance to agencies at all levels of government. It states that "the strategic objectives of homeland security in order of priority are to prevent terrorist attacks within the United States, reduce America's vulnerability to terrorism, and minimize the damage and recover from attacks that do occur." In support of this, it identifies information sharing as a foundation for achieving these objectives and explicitly cites the National Spatial Data Infrastructure (NSDI) as "a working example of compiling metadata to facilitate integration of data and support decision making." The NSDI (called for in Executive Order 12906,[15] issued by President Clinton in April 1994) is designed to be a network of federal, state, and local geospatial databases (NRC, 1993). The National Strategy for Homeland Security envisioned that the NSDI, as part of President Bush's e-government initiative, would include geospatial products, information, and enhanced metadata that would be coupled with incident management tools to allow real-time creation and display of maps and satellite images. Section 4.1 describes the current status of the NSDI as it pertains to emergency management.

3.3.2 Homeland Security Presidential Directive 7 (HSPD-7)

HSPD-7 establishes national policy on critical infrastructure identification, prioritization, and protection.[16] Section 31 directs the Secretary of Homeland Security to collaborate with other federal departments and agencies to develop a program to "geospatially map, image, analyze, and sort critical infrastructure and key resources by utilizing commercial satellite and airborne systems, and existing capabilities within other agencies." It allows for use of "national technical means," or classified imagery gathered by certain defense and intelligence agencies, as a last resort. In June 2006, DHS published a National Infrastructure Protection Plan (NIPP),[17] which says: "The Intelligence Community, the Department of

---

[14]*http://www.whitehouse.gov/homeland/book/.*
[15]*http://www.archives.gov/federal-register/executive-orders/pdf/12906.pdf.*
[16]*http://www.whitehouse.gov/news/releases/2003/12/20031217-5.html.*
[17]*http://www.dhs.gov/dhspublic/interapp/editorial/editorial_0827.xml.*

Defense, and other appropriate Federal departments, such as the Department of the Interior and DOT, are collaborating with DHS on the development and implementation of a geospatial program to map, image, analyze, and sort critical infrastructure/key resource (CI/KR) data using commercial satellite and airborne systems, as well as associated agency capabilities. DHS works with these Federal departments and agencies to identify and help protect those positioning, navigation, and timing services, such as global positioning systems (GPS), that are critical enablers for CI/KR sectors such as Banking and Finance and Telecommunications. DHS and the intelligence community also collaborate with other agencies, such as the EPA, that manage data addressed by geographic information systems." As part of this effort, it has developed a National Asset Database (NADB), which integrates geospatial resources, and is working on revising this. According to the NIPP, "The current NADB incorporates a flexible design to facilitate evolution, growth, and continued interconnectivity with additional databases and tools. Advancements will include integration with multiple commercial and Federal CI/KR databases, vulnerability assessment tools and libraries, intelligence and threat reporting databases, and geospatial tools into a single, integrated, Web-based portal."

3.3.3 National Response Plan

In December 2004, the Department of Homeland Security updated the National Response Plan to respond to objectives specified by the President in HSPD-5,[18] which directed DHS "to align Federal coordination structures, capabilities, and resources into a unified, all discipline, and all-hazards approach to domestic incident management." The NRP is the successor to (and supersedes) several other major plans, including earlier versions of the National Response Plan, the Domestic Terrorism Concept of Operations Plan, and the Federal Radiological Emergency Response Plan. The NRP adopts and adapts many of the functions and requirements detailed in those plans, including those related to geospatial data and tools. The NRP "is an all-hazards plan that provides the structure and mechanisms for national level policy and operational coordination for domestic incident management." It is designed to be partially or fully implemented either prospectively, in the presence of a threat or anticipation of a significant event, or retrospectively, in response to a significant event. The NRP also seeks to provide the framework for federal interac-

---

[18]*http://www.whitehouse.gov/news/releases/2003/02/20030228-9.html.*

tion and integration with other levels of government and other sectors with respect to all phases of disaster management. It describes federal capabilities, resources, and responsibilities, making reference to geospatial responsibilities in some cases. Importantly, the scope of the NRP is limited to federal departments and agencies that may provide assistance or conduct operations in actual disasters or other potential incidents of national significance, which it defines as "high-impact events" that require a coordinated response by a combination of federal, state, local, tribal, private-sector, and nongovernmental entities.

The NRP has five main components:[19]

1. The base plan, which "describes the structure and processes comprising a national approach to domestic incident management designed to integrate the efforts and resources of Federal, State, local, tribal, private-sector, and nongovernmental organizations";
2. Appendixes, which provide definitions and details;
3. Emergency support function (ESF) annexes, which "detail the missions, policies, structures, and responsibilities of Federal agencies for coordinating resource and programmatic support to States, tribes, and other Federal agencies or other jurisdictions and entities during Incidents of National Significance";
4. Support annexes, which describe administrative processes designed to facilitate plan implementation; and
5. Incident annexes, which address how the NRP is applied to particular contingencies or hazards.

Geospatial requirements appear in the ESF, support, and incident annexes as follows:

- *Emergency Support Function 5—Emergency Management Annex.* This annex specifies that DHS-FEMA, as the primary responsible agency, "coordinates the use of remote sensing and reconnaissance operations, activation and deployment of assessment personnel or teams, and Geographic Information System support needed for incident management." This annex also provides for the planning function in accordance with the NIMS. Specifically, ESF 5 "provides for the collection, evaluation, dissemination, and use of information regarding incident prevention and response actions and the status of resources." ESF 5 states that one of the functions of the planning section is to coordinate "with the DHS Science and Technol-

---

[19]At time of writing, the NRP was under revision, with an update expected in late 2006.

ogy Directorate and agencies with special technical capabilities to request support for geospatial intelligence, modeling, and forecasting."

- *Emergency Support Function 11—Agriculture and Natural Resources Annex* also makes reference to geospatial data. It specifies that with respect to the protection of natural and cultural resources and historic properties and to animal and plant disease pest response, the Department of Agriculture is responsible for coordinating with the Department of the Interior for relevant mapping and geospatial data and assessment tools. This annex also says that with respect to the safety and security of the food supply, the Department of Agriculture "provides Geographic Information Systems mapping capability for the meat, poultry, and egg product facilities it regulates to assist State and local authorities to establish food control zones to protect the public health."
- *Emergency Support Function 12—Energy* is intended to restore damaged energy systems and components during a potential or actual incident of national significance. Under Department of Energy leadership, "ESF #12 provides information concerning the energy restoration process such as projected schedules, percent completion of restoration, geographic information on the restoration, and other information as appropriate."
- *Emergency Support Function 13—Public Safety and Security Annex* provides for NASA geospatial modeling capabilities to be used as available.
- The *Tribal Relations Support Annex* specifies that federal departments and agencies are responsible for providing "appropriate incident management officials with access to current databases containing information on tribal resources, demographics, and geospatial information."
- The *Nuclear/Radiological Incident Annex* specifies that the Department of the Interior "advises and assists in the development of geographic information systems databases to be used in the analysis and assessment of contaminated areas, including personnel and equipment."

Overall, the committee believes that the NRP is weak in defining geospatial requirements and providing federal agencies specific direction about how to meet them, and that DHS's Geospatial Management Office should accelerate its plan to develop a geospatial concept of operations for the NRP.

3.3.4 National Incident Management System

The NIMS "provides a consistent doctrinal framework for incident management at all jurisdictional levels, regardless of the cause, size, or complexity of the incident." The NIMS provides a core set of doctrine, concepts, terminology, and organizational processes designed to promote effective, efficient, and collaborative incident management.

The NIMS rests on the well-established and broadly accepted ICS, which is employed to organize and unify multiple disciplines, jurisdictions, and responsibilities on-scene under one functional organization that directs incident operations. The ICS is classically organized into five major functions (command, operations, logistics, planning, and finance and administration; see Section 3.2.1). The NIMS suggests a sixth function, information and intelligence, which explicitly includes the analysis and sharing of geospatial data. The NIMS assigns responsibility to DHS's NIMS Integration Center (NIC) for facilitating the development of data standards for geospatial information, asserting that "the use of geospatial data must be tied to consistent standards because of the potential for coordinates to be transformed incorrectly or otherwise misapplied, causing inconspicuous, yet serious, errors." The NIMS sets forth the requirement that standards should be "robust enough to enable systems to be used in remote field locations, where telecommunications capabilities may not have sufficient bandwidth to handle large images or are limited in terms of computing hardware."

3.3.5 National Preparedness Goal

HSPD-8 focuses on strengthening preparedness,[20] and one of its requirements is the establishment of a national domestic all-hazards preparedness goal. This goal, developed with a capabilities-based planning approach habitually used by the military, is supposed to guide all levels of government, nongovernmental organizations, and the public "in determining how to most effectively and efficiently strengthen preparedness for terrorist attacks, major disasters, and other emergencies." An interim goal was published in March 2005 and "establishes the national vision and priorities that will guide our efforts as we set measurable readiness benchmarks and targets to strengthen the Nation's preparedness."[21] As of the October 2005 draft, neither the national preparedness goal nor the accompanying guidance that provides instructions on how to implement the goal addresses the nature or role of geospatial data or tools in preparedness.

According to the guidance, the goal is intended to be applied in conjunction with DHS's National Planning Scenarios and Target Capabilities

---

[20]*http://www.whitehouse.gov/news/releases/2003/12/20031217-6.html.*
[21]National Preparedness Goal (March 31, 2005); see *http://www.ojp.usdoj.gov/odp/assessments/hspd8.htm.*

List (TCL). The TCL identifies the capabilities required to perform the critical tasks identified in a Universal Task List (UTL), which provides a menu of tasks that may be performed in major events such as those illustrated by the National Planning Scenarios. Among these tasks, some are deemed critical. The UTL and TCL make only scant reference to geospatial data and tools, as follows:

- *Universal Task List.* Version 2.1 of the UTL was published in May 2005.[22] The UTL identifies approximately 1,600 tasks, of which 300 are deemed "critical." Critical tasks are defined as "those that must be performed during a major event to prevent occurrence, reduce loss of life or serious injuries, mitigate significant property damage, or are essential to the success of a homeland security mission." The UTL identifies one geospatially related critical task as part of the emergency management function: "Support identification and determination of potential hazards and threats including mapping, modeling, and forecasting." The UTL also identifies "common tasks" (i.e., tasks that cut across mission areas). One of the common tasks specified in the UTL is communications and information management, which includes "facilitate the development of geospatial information exchange standards" and "develop and maintain geographic information systems" as subtasks (neither of which is deemed critical).
- *Target Capabilities List.* Version 2.0 of this list was published in August 2006 and identified 37 target capabilities.[23] The TCL briefly references geospatial capabilities as relevant for four target capabilities: emergency operations center management, animal health emergency support, environmental health and vector control, and triage and pre-hospital treatment. In discussing resources needed for the management of emergency operations centers by cities, geographic information systems and geospatial imagery are listed as required resources to support planning. With respect to investigation of animal health emergencies, the TCL mentions that equipment must be able to enter, store, and retrieve geospatial information from the field and that geographic information systems may be used by epidemiologists to track the progress of an outbreak or to predict the impact of various management strategies.

Beyond the federal-level policy and doctrine described above, various other documents codify geospatial requirements. The DHS geospatial

---

[22]Universal Task List version 2.1, *http://www.ojp.usdoj.gov/odp/assessments/hspd8.htm*.

[23]Target Capabilities List version 1.1 is available at *http://www.ojp.usdoj.gov/odp/assessments/hspd8.htm*. A revised version was published in August 2006 and is available to the emergency response community, although it is not yet publicly available.

data model as mentioned earlier was just released in draft form during the writing of this report and therefore is not addressed in detail here. Other examples include national interagency plans and agency-specific plans, which are founded in either statutory or regulatory authorities and tend to pertain to specific contingencies. These plans provide protocols for managing incidents to be implemented by agencies that have jurisdiction, and often operate independent of DHS coordination and the NRP framework. Examples of such plans include, at the federal agency level, the National Oil and Hazardous Substances Pollution Contingency Plan; the Mass Migration Emergency Plan; the National Search and Rescue Plan; the National Infrastructure Protection Plan; and the National Maritime Security Plan.

While the focus in this section has been on DHS and policy initiatives or revisions since September 11, 2001, since these largely define the current disaster management operating environment at a national level, some other long-standing federal policies continue to impact emergency management. For example, in 2000, the Disaster Mitigation Act (DMA 2000, P.L. 106-390) amended the Robert T. Stafford Disaster Relief and Emergency Assistance Act (the legislation that enables FEMA to provide disaster assistance) to levy new mitigation planning requirements. Notably, however, DMA 2000 makes no mention of geospatial capabilities. In addition, FEMA's mitigation division manages several programs that focus on risk analysis and reduction. An important example is the National Flood Insurance Program, which makes federally guaranteed flood insurance available to citizens and businesses, promulgates floodplain management regulations to reduce damage, and identifies and maps the nation's floodplains.

Overall, direct reference to geospatial capabilities in federal policies is sparse. While there is general acknowledgement of the role that geospatial data and tools may play in incident response and management, no specific requirements are articulated in the National Response Plan or elsewhere. Further, there is no explicit reference to the role of geospatial data and tools in the pre-incident planning process. As a result, these policy documents offer little guidance or direction to governments in terms of the type or level of geospatial capability they ought to develop, or how these capabilities should be integrated into the broader emergency management architecture. They also provide little incentive for DHS to convene a robust team of geospatial experts that can be deployed rapidly to support field operations. The national disaster response could be significantly enhanced by integration and coordination of the various federal agencies' geospatial data capabilities and assets.

More importantly, certain needs articulated by the user community are unaddressed. One in particular deserves attention: the management and preservation of geospatial data related to major incidents. The National Response Plan specifies that FEMA is responsible for coordinating remote-sensing and geographic information system support. The NRP does not explicitly address the development and maintenance of data archives, however, and as a result such archives are rarely generated.[24] Geospatial data sets not only feed basic and applied research but are key to post-incident analysis that can inform future planning and preparedness activities. The time to develop data archives is during an incident; recreating them after the fact is well-nigh impossible.

3.4 GEOSPATIAL DATA NEEDS

Mission demands and the organizations that participate in fulfilling them vary across the phases of disaster and across hazard types. As a result, geospatial requirements also vary. Geospatial resources and processes must be able to adapt and respond to follow the contour of these changing demands. During the committee's deliberations, many individuals and agencies provided lists of the types of geospatial data most likely to be needed during the various phases of emergency response and associated tools and capabilities. Table 3.2 presents a summary of these discussions, showing some of the key user requirements and producer capabilities that were brought to the committee's attention, across the phases of emergency response. The table is not intended to be comprehensive, definitive, or prescriptive. It illustrates needs and current capabilities, highlighting some that are available versus some that are not. Its objective is to prompt further discussion about technology development and deployment.

From this table, some important general categories of user needs stand out. The most prominent requirements for geospatial data and analysis by decision makers are the following:

- Ability to assess risk and resilience;
- Pre-incident forecasts about hazard behavior, likely damage, property vulnerability, and potential victims;
- Decision aids to support recommendations for pre-positioning resources and evacuation;

---

[24]The committee was told that the U.S. Army Corps of Engineers and FEMA agreed following the 2005 hurricane season that FEMA will keep such an archive, and that the Corps will maintain an archive of Corps-developed disaster data.

- Timely, incident-specific locational information with respect to hazards, damage, victims, and resources, including information such as where people went, what kind of help is needed where, and the location of available resources;
- Ongoing monitoring of evolving hazards, response efforts, and resource status; and
- Insight into the interdependence and status of infrastructure components (energy, water, sanitation, road, communications, security systems, etc.) and awareness of critical infrastructure and facility vulnerability and status (refineries, chemical facilities, hazardous waste sites, bridges, tunnels, reservoirs, etc.).

3.5 CONCLUSION

This chapter begins with an elaboration of the processes and practices of emergency management and defines its key terms. Key elements of federal emergency management policy have been reviewed from the perspective of geospatial preparedness. Together, Chapters 2 and 3 provide the necessary background for Chapter 4, which presents a systematic review of the major themes underlying and impacting the integration of geospatial data and tools in emergency management, and lays out the committee's conclusions and recommendations.

TABLE 3.2 FOLLOWS

TABLE 3.2 Examples of Geospatial Needs and Capabilities

	Requirements
Mitigation	• Framework data, particularly detailed elevation data • Models, information, and analysis that can be used to develop grant guidance, analyze grant proposals, and assess plans • Data archive from previous incidents to support research and analysis • Research studies that can improve image analysis and inform resource pre-deployment and disaster response approaches • Improved understanding of changing environmental conditions post-disaster (e.g., new vegetation or flood maps) • Foundation data and imagery that allow for identification and graphic relationships among critical facilities, hazards, and resources • Clear understanding of infrastructure inventories, locations, relationships, and interdependencies • Risk and hazard maps • Ability to communicate with public about risk • Effective land-use planning using current local graphic information with incorporated hazards information and GIS decision support tools • Public, private, and nonprofit organization client databases • Improved understanding of the distribution of target populations at risk
Preparedness	• Critical infrastructure database (including information on high-risk occupancy facilities such as schools, medical facilities, and nursing homes) that includes attribute information • Foundation data and imagery that allow for identification and graphic relationships among critical facilities, hazards, and resources • Comprehensive geospatial database tied to full demographic profile for communities to yield understanding of populations at risk • Detailed geospatial data on the location and characteristics of businesses and the size of their workforce • Detailed geospatial data on the location and characteristics of equipment and supply assets as well as human assets • Identification of alternate sites for critical facilities • Pre-event imagery • Pre-plans that include building interior data • Database of current resource status and locations (e.g., shelters, vaccines, communications) • Shared parcel-level information (linked to tax assessor's or insurance industry data) • Spatial distribution and classification of residential structures by resiliency to hazards • Spatial distribution of social support need in at-risk communities • Standing annual contracts for geospatial capabilities • Sophisticated damage estimation models • Redundant data storage in geographically disparate locations

EMERGENCY MANAGEMENT FRAMEWORK

Current Capabilities	Gaps
• Digital elevation models developed from ground-based survey or processing of remote-sensing data—LIDAR, photogrammetry, or radar • Intelligent query of multiple spatial databases • Pre-event and post-event analysis (change detection) using remote-sensing and other geographic data • Geospatial analysis of project proposals in line with state policies • Visualization technologies that incorporate geographic risk data • Land-cover or land-use classification, change detection, and mapping using COTS GIS spatial analytical tools • Hazard models from government or commercial sources • Comprehensive geospatial database with full attribute data (may not be available in all communities)	• Modeling capability that determines and describes multiple effects due to dependencies in infrastructure and a single or multiple failures • Data to drive these models lacking in many communities • Robust, easily understood procedures that identify specific features of interest to emergency response managers in image data
• Critical infrastructure databases (where they exist) • Evacuation models and planning tools, and tools for monitoring traffic flow • Government and commercially developed framework mapping and standard COTS GIS products for mapping and spatial analysis • Image data from government programs such as the National Aerial Photography program, Google Earth, or commercial providers • Independent modeling of hazards impact • Land-cover classification for discriminating variation in residential structures using remote-sensing data supported by ground survey • Tools for tracking resource movement • Optimal location analysis capability in COTS GIS • Projected 24/7 population database that estimates population on 1 km grid resolution (ORNL Landscan population database—does not have age attributes)	• National cadastral database • National model or structure to share cost of database development • Comprehensive, current, accurate geographic database with census data and full attribute information for all features at the parcel level • A robust predictive model for estimating evacuation demographics—who will leave, where will they go, how long will they stay, who will come back—age is an important attribute • Incomplete up-to-date imagery (less than 3-5 years old) and detailed elevation data • Detailed geospatial data on the location and characteristics of equipment and supply assets as well as human resources

continued

TABLE 3.2 Continued

	Requirements
Response	- Ability to warn the public and notify responders - Ability to compare damage with client databases to calculate expected demand - Ability to track resource locations and status, including shelter sites - Ability to track the activities of public, private, and nonprofit service providers; maps of where current assistance is being provided - Rapid identification and categorization of the extent and type of damage over a widespread area, assessment of damage severity, including maps of damage areas and affected populations - Common operating picture based on shared geospatial data and analysis and continuous, real-time data about incident, damage, resources - Creation of an archive of social, economic, and geographic issues and responses for the incident - Detailed information on refugee and stranded demographics especially age and location and maps of needy and underserved areas - Robust communication system that supports data transmission from point of service to site of definitive analysis and decision making - Understanding of critical infrastructure damage (e.g., road and bridge closures, power outages) - Ability to provide coordinate locations for planning and executing search-and-rescue operations

EMERGENCY MANAGEMENT FRAMEWORK

Current Capabilities	Gaps
• Shared geospatial databases within individual cities and counties • State- or county-funded image acquisition • Visualization technologies • Application-specific remote-sensing data (i.e., multispectral data for environmental assessment or true-color, off-nadir high spatial resolution for structural assessments) with sophisticated image exploitation tool set • Coordinated access to government-developed response database • Correlation of individual-level data across data sets • Hazard model input to parcel-level geographic database for prediction of at-risk population • NOAA and FEMA Public Alert Warning System and reverse 911 • Residential structures damage estimation (RSDE) database • Robust geospatial analytical capability—COTS GIS, products for mapping and spatial analysis, and the ability to incorporate model output • Sophisticated, nearly incident-specific, remote-sensing, image acquisition, and exploitation capabilities • The ability to geo-code coordinates to support search-and-rescue operations	• Rapid dissemination of maps with hazard and victim locations to responders • Capability to track the location and characteristics of equipment and supply assets as well as human assets • Fleet tracking systems that provide full resource location in a dynamic context (possible but not used) • SOPs for remote-sensing and GIS technologies in emergency management agencies • Integrated system for real-time reception of remote-sensing data to forward deployed capability • Coupled modeling capability to spatial decision support system or simple GIS • RSDE system is not integrated with GIS database for real-time automated update • Integrated, location-based field data acquisition system linked to central GIS for use by initial response teams and recovery teams • Dynamic update of geospatial database content from any approved point in the response activity • Assured communication system for geography-specific public alert and feedback from affected population on status and need • Coordinated, detailed information on post-incident population movement • Rapid damage assessment identifying extent and severity of damage

continued

TABLE 3.2 Continued

	Requirements
Recovery	• Ability to provide information to public about rebuilding and regrowth • Ability to track resource locations and status, and the locations and activities of service providers • Access to response geospatial database for transition of response to recovery • Geospatial tools for land-use planning • Identification and analysis of optimal landfill, shelter, long-term housing sites, disaster recovery centers, and recovery team staging areas • Integrated monitoring system for recovery operations at the parcel level • Maps of how population shifts as a result of disaster—age is an important attribute • New information required to issue building permits • Remote-sensing acquisitions to monitor recovery progress on a regional basis • User-friendly decision support tools to systematically evaluate short- and long-term demands such as allocation of resources, capacity shortfalls, and status of restoration

NOTE: COTS = commercial, off the shelf; ORNL = Oak Ridge National Laboratory; SOP = Standard Operating Procedure.

Current Capabilities	Gaps
<ul><li>Optimal location analysis using image data, geographic data, and spatial modeling</li><li>COTS GIS tools for spatial analysis of optimal siting and land-use planning (e.g., landfill, shelter)</li><li>Commercial or government-provided remote-sensing acquisitions to monitor recovery progress on a regional basis</li><li>Land-cover or land-use classification, change detection, and mapping using COTS image analysis tools</li><li>Correlation of individual-level data across data sets</li><li>Multiple overlay and spatial relationships and comparison</li><li>Standard COTS GIS products for mapping and spatial analysis (but data may not be available)</li></ul>	<ul><li>Fleet tracking or location-based service to tag field activity with a handheld device; used by private sector (e.g., FedEx) but not by FEMA</li><li>Dynamic models that incorporate real-time geographic data of response activity within a GIS for full understanding of resource use and changing need</li><li>Coordinated, detailed information on post-incident population movement</li><li>Simple geocoding capabilities that allows nontechnical staff to provide coordinates for search and rescue operations</li></ul>

4

The Challenge: Providing Geospatial Data, Tools, and Information Where and When They Are Needed

In this era of heightened requirements for prompt and effective response, rapid access to disparate geospatial information sources is essential. As shown in Chapters 2 and 3, the emergency management community relies heavily on the ability to discover and use accurate up-to-date information in order to respond to disasters and other emergency events. However, the necessary data are scattered among numerous agencies, there are many impediments to rapid access, the skilled personnel needed to work with the data and tools are often not available in sufficient quantity, and the technological environment is changing constantly, causing endless confusion. This chapter explores these and other related issues in greater depth. Each section of the chapter takes one issue, describes the problem in detail, elaborates on its significance, describes possible solutions, and where appropriate, offers recommendations. This overview and the first three sections deal with issues that require policy changes; the next three focus on operational changes that could be made to enhance the use of geospatial data and tools; the next two sections on tools and training discuss changes that will produce better utilization in the future; and the final section addresses funding.

It is important to note that this study deals with the intersection of two distinct communities—the emergency response community and the geospatial community. The issues discussed may have their roots in one community or the other, but the resolution of these challenges will require both communities to work together, as reflected in the recommendations. The fact that both of these are professions in their own right, with the emergency management community often seen as conservative with regard to the adoption of new technologies, presents a challenge. Without the support—and preferably the leadership—of the emergency management community, the geospatial data community's own efforts will have little benefit.

The committee heard from many federal, state, and local emergency management professionals during its deliberations and during the study's workshop, as well as from several representatives of the private sector and nongovernmental organizations (NGOs). All testified to the central importance of geospatial information. The first questions responders ask when a disaster occurs are, Where is it? Where are the victims? Where are the hazards? Where are the resources? The first request from an incident commander is often for a map, and the need is immediate. Responders must act within a "golden hour," during which delivering victims to appropriate care providers has the best chance of saving lives.

Data on the cost savings from more effective emergency management are almost impossible to compile, in part because many benefits, such as lives saved, are impossible to value and in part because any form of controlled experiment in which costs are compared with and without effective emergency management is impossible to conduct. Nevertheless some of the more direct cost savings might be quantified, in certain limited contexts. For example, The National Governors Association Center for Best Practices published an Issue Brief on State Strategies for Using IT for an All-Hazards Approach to Homeland Security (July 13, 2006).[1] In the section about geographic information systems (GIS), it has the following paragraph:

> State and local governments in Virginia combined their efforts in October 2001 to launch the Virginia Base Mapping Program (VBMP) for use in deploying resources and personnel during disasters. At an estimated cost of $8.2 million, this program began delivering DVDs [digital video discs] with GIS technology to 134 cities and counties in February 2003, providing information about transportation systems, private-sector facilities, natural resources, and many other assets. Although measures of lives saved, injuries averted, and property damage avoided are difficult to calculate, it is estimated that in its first year the VBMP saved the state between $5 million and $8 million in operating costs.

Responders and managers need to be able to work with several map layers or themes. The most important layer to them is the search grid, which must be established quickly and applied by all agencies working

---

[1] http://www.nga.org/Files/pdf/0607HOMELANDIT.PDF.

THE CHALLENGE 89

on the incident. They also need to be able to locate points on the map and on the ground. While street address normally provides an easy way to do this in urban areas, it is often unsatisfactory in rural areas or when street signs and house numbers have been obliterated. The Global Positioning System (GPS) provides an effective and universal alternative, but requires that maps be overprinted with GPS coordinates, using latitude-longitude, Universal Transverse Mercator coordinates, or the proposed National Grid (the National Grid was endorsed and adopted by the Federal Geographic Data Committee in 2001).[2] Further, they need to be able to map an event as it changes in real time and to print and distribute updates quickly. From an emergency management perspective, maps enable the location-specific assessment of hazard, risk, vulnerability, and damage. They are required with different levels of geographic detail throughout the emergency management cycle, from the moment an incident occurs through long-term recovery and into mitigation.

For most emergency events, the needed geospatial information and services for planning and response are maintained by a variety of public and private organizations in multiple jurisdictions. Government agencies are stewards of large volumes of data, most of which are held by state or local agencies. However, additional key layers, such as critical infrastructure data, are maintained by the private sector.[3] As mentioned previously, estimates from the Department of Homeland Security's Protected Critical Infrastructure Information (PCII) Program are that the private sector owns and operates 85 percent of the nation's critical infrastructure.[4] Many of these organizations are members of local utility notification centers, also referred to as "One Call" or "Call Before You Dig" agencies. However, data are shared with these and other consortia under very restrictive agreements and may not be used for any other purpose, even during emergencies.

Emergency preparedness and response require data from many sources both public and private, and critical infrastructure information is but one of many themes that must be accessed. There are also needs for property records, street centerlines, floodplain delineations, and other data that are maintained by the public sector. From an emergency preparedness and response perspective, it is critically important for all sources of data to be utilized to ensure that planners and responders have

---

[2]*http://www.fgdc.gov/standards/projects/FGDC-standards-projects/usng/index_html.*
[3]*http://www.dhs.gov/dhspublic/display?theme=92&content=3760/.*
[4]See PCII Program overview at *http://www.dhs.gov/dhspublic/interapp/editorial/editorial_0465.xml.*

the best possible common understanding of the operating picture. However, although many of the data that are needed by emergency managers are already developed by other organizations for other purposes in the general course of local government and community development, various issues and challenges prevent easy access to or use of these data for emergency management.

Data on the ownership of land parcels, or cadastral data, provide a particular and in some ways extreme example of the problems that currently pervade the use of geospatial data in emergency management. Vast amounts of such data exist, but they are distributed among tens of thousands of local governments, many of which have not invested in digital systems and instead maintain their land-parcel data in paper form. As with many other data types, it is not so much the existence of data that is the problem, as it is the issues associated with rapid access. In their report *Parcel Data and Wildland Fire Management*, Stage et al. (2005) argue that cadastral data can provide the most current and accurate information in support of emergency management, but note that access to such information can be limited by a number of factors including the following:

1. Data distribution agreements. In some cases, local units charge for the data or have data licensing agreements that constrict access to the information.
2. Data format. The data might be in a format that is not recognized or usable by responding agencies.

These and other issues identified in Chapters 2 and 3 are explored in depth in subsequent sections of this chapter.

Local emergency responders generally have vast personal knowledge of their communities, and as a result the use of geospatial information may sometimes be seen as superfluous to their immediate needs. However, when disasters extend far beyond the boundaries of a community, when local responders are unable to respond adequately and professionals without knowledge of the area must be brought in from elsewhere, or when impacts extend to infrastructure such as underground pipes about which local responders have little personal knowledge, then geospatial data and tools become absolutely indispensable to an effective, coordinated response.

Conclusion

As the committee heard in testimony, geospatial data and tools are essential to all aspects of planning for disaster and to all aspects of community resilience. In this respect, the committee echoes a conclusion of an

THE CHALLENGE

earlier National Research Council (NRC) study: "Much of the information that underpins emergency preparedness, response, recovery, and mitigation is geospatial in nature" (NRC, 2003, p. 1). Without knowledge of where the event has occurred, the area it has impacted, the nature of impact in each part of that area, or the locations of shelters and potential responders, and without access to the tools to analyze such information and to present and distribute it in useful form, the eventual impact of the event will necessarily be greater than it need be, whether measured in loss of life, injury, damage to property, or disruption of essential activities. Also, although many of the geospatial data needed for emergency response generally have already been developed by communities for other purposes, there are a variety of issues that currently impede their use for emergency management. Therefore, steps must be taken to explicitly recognize and meet the geospatial needs of the emergency management field. As its first, overarching conclusion, the committee believes that the importance of geospatial data and tools should be recognized and integrated into all phases of emergency management and, specifically, into the national plans and policies reviewed in Section 3.3 and existing emergency management procedures.

> **RECOMMENDATION 1:** The role of geospatial data and tools should be addressed explicitly by the responsible agency in strategic planning documents at all levels, including the National Response Plan, the National Incident Management System, the Target Capabilities List, and other pertinent plans, procedures, and policies (including future Homeland Security Presidential Directives). Geospatial procedures and plans developed for all but the smallest of emergencies should be multiagency, involving all local, state, and federal agencies and NGOs that might participate in such events.

4.1 FOCUS ON COLLABORATION

The lack of consistent policy for collaboration, together with protocols and structures for coordination and communication, has long been an impediment to effective collaboration, sharing, and reuse of geospatial data and tools among all levels of government. Since the early 1990s a number of government initiatives and orders have charged federal agencies with coordinating their programs in this specific area.

In 1990 the Federal Geographic Data Committee[5] (FGDC) was formed and given the lead responsibility for this coordination by an updated Of-

---

[5]*http://www.fgdc.gov.*

fice of Management and Budget (OMB) Circular A-16.[6] In 1994 the FGDC was also charged by Executive Order 12906 to provide leadership in coordinating the federal government's development of the National Spatial Data Infrastructure (NSDI) and to seek the involvement of other levels of government and sectors in this endeavor.[7] Federal-level coordination has produced benefits in the development of more than 20 standards supporting the NSDI, the implementation of the NSDI Clearinghouse Network,[8] the Geospatial One-Stop,[9] and the emerging Geospatial Profile for the Federal Enterprise Architecture.[10]

State-level coordination has also produced many improvements. The National States Geographic Information Council (NSGIC) has been an effective mechanism for facilitating coordination among states.[11] NSGIC's activities have leveraged the strong geospatial programs present in a number of states to bring about improvement of coordination activities in many other states. Private-sector and professional organizations have also played important roles in facilitating coordination among various segments of the geospatial community and have likewise produced benefits for participants. However, these efforts have been confined primarily to local jurisdictions and, as such, have proven difficult to replicate across a wider spectrum.

Specific examples of effective collaboration exist in many places both across the nation and internationally. There are excellent resources already available that describe the issues involved in collaboration and suggest approaches to enhancing cooperation across jurisdictions. One such project developed by the Geospatial Information and Technology Association (GITA) is entitled GECCo (Geospatially Enabling Community Collaboration).[12] Another resource is the work of the Open Geospatial Consortium as part of the Critical Infrastructure Protection Initiative (CIPI) completed in 2002 and 2003,[13] and another is the work done by Emergency Management Alberta.[14] In these three examples, a common principle is that agreements must be discussed, negotiated, and formalized

---

[6]The current version of the circular can be found at *http://www.whitehouse.gov/omb/circulars/a016/a016_rev.html*.
[7]*http://www.archives.gov/federal-register/executive-orders/pdf/12906.pdf*.
[8]*http://www.fgdc.gov/dataandservices/*.
[9]*http://gos2.geodata.gov/wps/portal/gos/*.
[10]*http://www.fgdc.gov/fgdc-news/geoprofile20050131/*.
[11]*http://www.nsgic.org*.
[12]*http://www.gita.org/ngi4cip/gecco.pdf*.
[13]*http://www.opengeospatial.org/initiatives/?iid=64/*.
[14]*http://www.municipalaffairs.gov.ab.ca/ema_index.htm*.

before an emergency situation occurs if the impacts of institutional and social barriers to interoperability are to be reduced.

Many types of agreements are needed, including the following:

- Data-sharing agreements among public and private organizations
- Proprietary agreements so that geospatial data can be used during emergencies without becoming part of the public domain
- A predefined list of geospatial and other technical personnel and vendors required in support of a response to an event
- Guidelines for sharing data with the media during and after an event
- Agreement on interoperability standards to enable the on-demand access, integration, and exchange of relevant geospatial data
- A process to organize, integrate, and distribute both data internal to an organization and data from other organizations

These agreements can take considerable time and energy to put in place, but if they are not, the results can be at a minimum very frustrating and at worst devastating. However, despite efforts at various levels and within sectors, collaboration between levels of government and with other sectors has been difficult to achieve. The FGDC has been seeking to carry out this role for the geospatial community; however it has not achieved complete success due to lack of authority, budget, and resources.

The FGDC's Future Directions Initiative recently provided a high-level look at the nation's sharing and use of geospatial information and the development of the NSDI.[15] The study report finds that geospatial data and information have been identified as valuable assets in conducting the business of government. In the post-9/11 era, there is a heightened appreciation of the importance of geospatial data to support homeland security needs and other critical requirements. There is a clear sense of urgency that the problems associated with intergovernmental and intersector collaboration in geospatial data production, access, and sharing need to be resolved in a timely and comprehensive manner.

The Future Directions Initiative study team found widespread agreement that the NSDI requires strong national leadership, that all sectors should be represented in the leadership and governance process, that stable funding and political support are required, and that an effective NSDI requires a clear national strategy to complete and maintain the framework layers. The team found a broad consensus that a strong and

---

[15]"Future Directions—Governance of the National Spatial Data Infrastructure," final draft report of the NSDI Future Directions Governance Action Team, May 31, 2005.

renewed national focus is needed to drive our country toward the production of highly accessible, accurate, and reliable geospatial data. The team believed that a national approach, incorporating all sectors, is necessary to accelerate the production of geospatial data for the NSDI and to ensure its ongoing maintenance. The increasing ubiquity of geospatial data and tools lends urgency to the need for current, complete, accurate, and nationally consistent data. The study team recommended the establishment of a new governance structure to provide national leadership in the development of the NSDI, with participation from multiple sectors.

The Committee on Planning for Catastrophe also reviewed the current governance structure of the NSDI in light of this study and discussed whether it was adequate to provide effective coordination across state, local, and federal governments and the private and not-for-profit sectors in the particular context of emergency management. The arguments and conclusions of the Future Directions Initiatives study resonated strongly with the committee, which concluded that the proposed changes in the governance structure would provide a much more effective framework for geospatial data and tools in emergency management. Moreover, the committee felt that it was desirable for the needs of emergency management to be addressed within this larger framework and that the emergency management community should be given a sufficiently strong voice to ensure that these needs are met.

Conclusion

A national geospatial governance process such as the one described above would do much to improve the attention given to policy and other institutional issues that make it difficult for the different levels of government and other sectors to work together effectively in the development of geospatial capabilities for emergency management. Whatever the root cause of a disaster—terrorism, natural occurrences, or accident—the methods of preparing for, responding to, recovering from, and mitigating the effects of such events, and ideally preventing reoccurrences, are based on a common approach: the collaborative and coordinated use of geospatial data and tools. This cannot happen without the many mutually dependent agencies and organizations charged with protecting our nation's citizens and infrastructure being able to share their geospatial data and tools efficiently and effectively for emergency management purposes. Moreover the special circumstances of emergency management—the need for speed and for planning in advance without knowledge of where and when disaster will strike, and the extreme costs in damage and loss of life that may result from a bungled response—all give additional merit to arguments in favor of greater collaboration and effective governance.

THE CHALLENGE

The myriad of individual and organizational collaboration efforts are currently doing much to resolve specific local needs and to provide a positive, dynamic environment for collaboration. Many problems and issues remain, however, and many of these successful efforts have been costly in terms of the time required to develop and maintain them. Missing is a strong, nationally focused governance process to bring the relevant and affected organizations together within the established framework of the NSDI to ensure collaborative approaches to resolving multijurisdictional and national-level issues. The kind of governance process described by the report of the FGDC Future Directions Initiative is the subject of continued discussion within the NSDI community and could significantly improve the environment for collaboration and data sharing during emergency response. The Department of Homeland Security (DHS) has been assigned responsibilities for coordinating geospatial data and tools for emergency management, as detailed in Section 3.2.2. The committee therefore recommends that DHS play a leading role in ensuring that this proposed strengthening of NSDI governance addresses the needs of emergency management.

RECOMMENDATION 2: The current system of governance of the NSDI should be strengthened to include the full range of agencies, governments, and sectors that share geospatial data and tools, in order to provide strong national leadership. DHS should play a lead role in ensuring that the special needs of emergency management for effective data sharing and collaboration are recognized as an important area of emphasis for this new governance structure.

4.2 GEOSPATIAL DATA ACCESSIBILITY

A critical requirement for emergency preparedness, response, and mitigation is to have rapid access to the most accurate, up-to-date geospatial content, whether it be current wind speed and direction, the location of hospitals, damage assessment data, or the results of predictive flood models. Emergency managers and responders need rapid and reliable access to such content on demand. However, there are numerous issues involved in meeting the challenges of this on-demand, rapid-access requirement. Whether the geospatial data are being accessed from archives or from real-time sensor feeds, the following must always be considered if we are to build a national asset not just for emergency management but also for other homeland security functions:

- Are geospatial data being collected once and maintained by the organization that can do this most effectively?

- Is it possible to combine geospatial data seamlessly from different sources and to share them between many users and emergency applications?
- Are geospatial data available for use in emergency management, or do use conditions restrict their availability?
- Is it easy to discover which geospatial data are available, to evaluate their fitness for the purpose, and to know which conditions apply to their use?

There are both policy and technology impediments to the achievement of these goals. Some of the issues deal with sharing data among organizations, since there are many reasons why a data-producing agency may be reluctant to make its data available, such as concerns related to privacy, confidentiality, or liability. Other issues are more technical in nature and are focused on the interoperability of data and the need for standards that address not only the content and labeling of data, but also real-time discovery of and access to data through clearinghouses and portals. Finally, although data may be accessible, there may be questions related to their quality. This section describes these various challenges.

4.2.1 Data Sharing

The unwillingness to share geospatial data is by no means universal, and many entities make their data free and easily accessible for use by the public. Many do not, however, particularly local governments or private utility companies, where some of the most important geospatial data for emergency management often reside. There are a number of reasons for this reluctance to share data, including the following:

- The desire to sell data to obtain revenue from a costly and valuable asset;
- The considerable effort required to convert data into a form in which they can easily be shared, especially at the local level;
- The fear that data may assist terrorists in their activities;
- A basic distrust of the entity requesting the data or a basic unwillingness to cooperate;
- A concern for liability if the data are improperly used or are of insufficient quality for a specific use;
- The fear that once others are aware of the existence of data they may attempt to obtain access to them through freedom-of-information laws; and
- The most basic fear that the organization will lose control of its data.

THE CHALLENGE   97

A workshop panelist told the committee that government agencies were much more willing to share data for homeland security than for other purposes, but were adamant in many cases that access be restricted to that purpose. In some cases, agencies went so far as to agree to forward certain data that they deemed sensitive only after an incident had occurred.

Currently, policies for geospatial data sharing within specific levels of government are set by their executive branches. The policies developed for each level of government vary, and enforcement varies within each level from department to department (Sidebars 4.1 and 4.2 contrast two different approaches to policy formulation). Almost none of the policies set for one level of government are imposed on another level, and many local governments have no policies for sharing geospatial data at all. It is impractical to expect that the data-sharing policies of all government entities will be the same. However, it is reasonable to expect that all government and private entities have clearly defined data-sharing policies and guidelines, especially for data relevant to emergency management.

As the committee heard in testimony from many individuals and agencies, the lack of such policies and guidelines results in confusion for data custodians and becomes a nightmare for those wishing to acquire data on a large-scale basis either before or during an event. A significant amount of time and staff effort is required to investigate each data owner's issues and policies and to keep abreast of changes in these policies. For example, New York State has been aggressive in collecting and maintaining geospatial data since the events of September 11, 2001. A representative of New York State's Office of Cyber Security and Critical Infrastructure Coordination reported to the committee that it has had a team assigned to geospatial data collection and maintenance for homeland security and emergency response since 2002 and has collected more than 850 sets of geospatial data. However, it was noted that this involved significant effort because each government entity required personal contact to discuss how its data sets would be used and where they would be stored. In one case the office had been unsuccessful in negotiating for certain utility data from a federal agency. In another case, more than two years of effort were required to obtain the use of a local government's parcel data. Data produced by federally funded research and development centers (FFRDCs) and GOCO (government-owned, contractor-operated) national assets are often not included in data-sharing agreements between government agencies. This presents additional barriers to effective data sharing.

Utility companies, in particular, have been viewed as organizations that create and compile sophisticated databases but will not readily share these data with others except in emergency situations. The first concern of

Sidebar 4.1
The Case for Mandatory Data Sharing

The National Pipeline Mapping System
As a means of creating a single source of information, the U.S. Department of Transportation (USDOT) requires all transmission pipeline operators to provide data on an annual basis to the National Pipeline Mapping System (NPMS).[a] Section 15 of the Pipeline Safety Improvement Act of 2002 required that operators provide "geospatial data appropriate for use in the National Pipeline Mapping System or data in a format that can be readily converted to geospatial data," together with other information on pipeline operations. The NPMS enforces strict mapping and metadata standards that must be followed by all operators. Its intent is to provide a common national database depicting the location of all pipeline networks and related attribute information that can be accessed whenever needed.

The terrorist attacks of September 11, 2001, placed additional security concerns on the U.S. pipeline infrastructure; therefore, access to the NPMS is restricted to federal, state, and local government agencies (including emergency responders). Participating pipeline operators are able to view only their own data on-line from the USDOT's web site and cannot view data from any other operator. Operators who do not comply by the annual mid-June deadline are subject to a minimum $1 million penalty that increases the longer the operator is noncompliant. This action, however drastic it may seem, is one example of how data sharing can be made a mandatory rather than a voluntary venture.

The NPMS standards are a good example of program-oriented specifications from a federal agency, providing clear guidance on how to prepare and submit data. If they could be extended to bring them into line with the more general geospatial data standards of the International Organization for Standardization (ISO),[b] the Open Geospatial Consortium (OGC),[c] and the FGDC,[d] it would be easier to integrate NPMS data with other geospatial data in emergency management operations as part of a national fabric of critical infrastructure information within the framework of the NSDI.

[a]http://www.npms.rspa.dot.gov.
[b]http://www.iso.org.
[c]http://www.opengeospatial.org.
[d]http://www.fgdc.gov.

Sidebar 4.2
The Case for Voluntary Data Sharing

The 50 States Initiative

This project seeks to support the NSDI by coordinating data-sharing methodologies and standards at the local level, providing a common framework in which all states can participate.

The effort is coordinated by the National States Geographic Information Council (NSGIC), and plans to establish 50 state coordinating councils that will contribute routinely to the governance of the NSDI.

Statewide councils will bring consistency to the NSDI by

- Serving as a focal point to aggregate the activities of all sectors into the NSDI in a functional way;
- Providing incentives for non-federal entities to adopt appropriate national standards;
- Working together on data production, infrastructure, and application development to avoid duplication;
- Ensuring routine data access by all sectors;
- Establishing sharing agreements for data not in the public domain;
- Publishing lists of data stewards and integrators for framework themes;
- Publishing metadata for framework themes in the NSDI clearinghouse;
- Providing functioning tools for maintaining the clearinghouse inventory;
- Participating in the National Map initiative;[a] and
- Adopting appropriate data-sharing standards including

— A commitment to implement appropriate Open Geospatial Consortium (OGC), FGDC, American National Standards Institute (ANSI),[b] and International Organization for Standardization (ISO) standards;

— Posting local, state, and tribal framework data to the clearinghouse or otherwise making them available through interoperable interfaces; and

— Promoting the adoption and incorporation of appropriate OGC, FGDC, ANSI, and ISO standards and interoperable practices among local, state, and tribal agencies.

---

[a]http://nationalmap.gov.
[b]http://www.ansi.org.

many utility companies is that their data will become part of the public domain and be available to anyone. From a utility perspective, this has serious liability issues. For example, if an individual or organization wanted to perform some type of activity that required excavation and utilized information about underground facilities that had been made available to the general public but proved to be dated and therefore inaccurate, the utility company could be liable for any damages incurred. The risk of this happening is very real, especially if the proper method of submitting a facility location request with the local utility notification center within 48 hours prior to performing any excavation tasks was not followed. Therefore, adequate measures to ensure that these data are excluded from the Freedom of Information Act (FOIA) are a basic requirement. Second, to address the need to accurately track facility information, utility companies have spent a great deal of time and money to build sophisticated databases, which are consequently seen as corporate assets and treated as such. Therefore, the prospect of sharing these data without the protection of a proprietary agreement is often not an option. These two barriers must be addressed if data-sharing and collaboration initiatives are to prosper, perhaps through the creation of suitable incentives for data owners to participate in these types of activities.

However, it is important to note that data sharing does not necessarily mean sharing one's data in their entirety, but rather can be limited to key data elements (e.g., commodity, location, size, material, ownership—to name a few). Portions of an organization's data may be proprietary, confidential, or sensitive or may require protection as intellectual property, whereas other portions may be suitable for limited or full data sharing. Where there are legitimate reasons to protect portions of a data set, identifying the critical data elements that are relevant in responding to and planning for emergency events is particularly appropriate. Organizations vary in the level of geographic detail of their databases and in the complexity of the attributes that are maintained to meet regulatory and internal needs. Only a subset of attributes may be needed for emergency management, and lower levels of geographic detail may also be sufficient. Data sharing may be much more palatable to data owners if it involves only subsets of attributes or coarser levels of geographic detail.

4.2.2 Interoperability

As defined in Section 1.3.2, interoperability is about the ability of two or more systems to share data and tools effectively and seamlessly, independent of location, data models, technology platform, terminologies, and so forth. To achieve true interoperability (or the effective sharing of geospatial data and tools), many levels and aspects of interoperability must be understood. The most obvious, and perhaps the easiest to solve, is technical interoperability, which is concerned primarily with issues of format. More problematic is semantic interoperability, which addresses and overcomes differences in concepts and the meaning given to data by different users and systems. These core semantic differences are reflected in the selection and definitions of technical terms used in publications, communications, and databases. (Institutional, human, and political issues that make it difficult for individuals and organizations to work together and lack of legal interoperability due to inconsistencies between the legal contexts in which different individuals and organizations operate are discussed in other sections of this chapter.) Achieving effective interoperability for emergency management may require radical changes to the ways in which organizations work, especially to their attitudes about information. The following material expands on each of these sources of interoperability problems.

Technical

This is the "nuts and bolts" of software and hardware interoperability, where the work of standards organizations can best be leveraged. Technical interoperability is typically achieved by selecting and implementing the appropriate software and Internet standards, common content encodings for transmission, and so forth. Within an enterprise, technical interoperability is quite often the easiest form of interoperability to achieve for any given business process. Yet, in cases where different data formats are encountered in the field during response, it can still cause significant delays in developing useful geospatial products.

Semantic

Geospatial data represent an extremely rich conceptual domain that requires special attention, perhaps more so than any other type of data. The enormous variety of ways of encoding geospatial data and the large number of classification schemes, vocabularies, terms, thesauri, and data definitions in use by data-producing agencies make it particularly challenging to process requests for geospatial information. Within any organization seeking to integrate geospatial data, it is vitally important that there is agreement on the proper definition and use of metadata. In principle, proper metadata can provide the foundation for semantic interoperability, by defining the meaning of each of the terms that underlie the data production process. In reality, however, the difficulties of overcoming differences of culture, language, and discipline may be far beyond the capacity of current metadata standards and practices. Efforts to map terminologies across domains have proved challenging, but solutions to this challenge are critical to ensuring accuracy and efficiency in data sharing.

Issues related to interoperability are often addressed through establishment of standards. For example, the issue of communication of data is being addressed by the Organization for the Advancement of Structured Information Standards (OASIS), which is creating the Emergency Distribution eXchange Language (EDXL) standard that will be incorporated in the National Information Exchange Model (NIEM). OASIS is a well-recognized standards organization that works on e-business standards—many of which are required in the emergency management world. Therefore, there is an Emergency Management Technical Committee that has representatives from many organizations, such as DHS, Warning Systems Inc., the Capital Wireless Integrated Network (CAPWIN), the Open Geospatial Consortium (OGC), Environmental Systems Research Institute (ESRI), and many others. Initially, EDXL was an outgrowth of work at the Department of Justice and will be a multipart standard. The initial part that has been approved by OASIS membership is EDXL-DE, where DE stands for Distribution Element. The primary purpose of the Distribution Element is to facilitate the routing of any properly formatted Extensible Markup Language (XML) emergency message to recipients. The Distribution Element may be thought of as a "container." It provides the information to route "payload" message sets (such as alerts or resource messages) by including key routing information such as distribution type, geography, incident, and sender-recipient IDs. There are very few implementations of this new standard as yet.

There is also a need for standards that address not only the content or communications of data, but also the real-time discovery of and access to data through clearinghouses and portals. Part of this solution is agreement on best practices regarding the standards of content and service interfaces that can best help achieve these goals and meet the requirement of providing geospatial content to emergency managers and personnel when and where it is needed.

Examples of how other countries are defining best practices for on-demand access to geospatial data can be seen in various e-government and spatial data infrastructure activities in the European Community,[16] the United Kingdom,[17] Germany,[18] and New Zealand.[19] All of these

---

[16] *http://inspire.jrc.it.*
[17] *http://www.govtalk.gov.uk/schemasstandards/egif.asp.*
[18] *http://www.kbst.bund.de/SAGA-,182/start.htm.*
[19] *http://www.e.govt.nz/standards/e-gif/.*

activities include a variety of geospatial standards using consistent, standards-based implementation architectures.

4.2.3 Data Quality

The committee heard many comments to the effect that use of the most accurate and up-to-date framework and foundation data is essential to successful response and recovery. These data almost always reside at the local government level or with the private sector; at the local government level, they are generally the by-product of routine business processes such as the creation of parcel data for property taxation. Because they serve everyday business needs, they are kept up-to-date and represent the most accurate data available. Furthermore, local governments, which are usually the first responders to emergencies, are the most familiar with these data and understand their strengths and weaknesses (the importance of education in issues of geospatial data quality and uncertainty is addressed in Section 4.8).

The use of geospatial data during the 9/11 recovery efforts was much more effective because the local geospatial professionals working with these data were familiar with them and understood their complexities and high level of accuracy. The committee heard that when federal agencies arrived on-site with their less accurate national data, they realized the benefits of the more accurate local data and converted to them after several days to improve overall coordination. During large disasters, the use of the same framework and foundation data by responders in various parts of the country is vital for close coordination.

Some municipalities and states have aggressively pursued the gathering and improvement of data needed to respond to emergencies, but many others have not. Unfortunately, instead of coordinated multiagency efforts to organize data, assist in improving their accuracy, and create new data to fill gaps, the committee heard about many redundant efforts to gather, recreate, or purchase similar data at various levels of government. This results in significant amounts of funding being needlessly spent and time and effort wasted. As a result, an opportunity to focus on improving the original local government data and reusing them through all levels of government is often missed.

Furthermore, geospatial information is only as good as the available data. The lack of established quality assurance and control (QA/QC) standards and testing processes results in data inconsistency and inaccuracies, which can negatively impact analyses used by decision makers. Determining whether differences in analyses and recommendations are the result of inconsistent or inaccurate data is extremely time-consuming and may be difficult to accomplish in a time-critical response environment.

Conclusion

As Sections 4.2.1-4.2.3 have shown, current arrangements for geospatial data access in support of emergency management resemble a complex patchwork that is time-consuming to establish and maintain and confusing for all involved. There is little agreement or consistency in such technical issues as the formats that will be used, the locations where data will be made available, the security mechanisms that will prevent unauthorized access, or the architecture of the servers. This patchwork makes it difficult for agencies to acquire the data needed to prepare for, respond to, recover from, and mitigate the effects of disasters; makes it difficult to identify gaps in data coverage or to address problems related to data quality; and is one factor among many in determining eventual costs, injuries, and possibly even loss of life. With some exceptions, there are both confusion in the legislative context and an acute lack of consistency across levels of government.

While a few data custodians prefer to not provide their data until after an event has occurred, presentations to the committee made it clear that access is needed prior to an event during the preparedness phase in support of training and planning, and because pressures during the response phase are so great that the process of acquiring data would delay other essential activities.

The level of interoperability necessary to enable systems to exchange and use data and tools, without special effort on the part of the user, can essentially be achieved by the adoption of policies stressing the importance of interoperability and requiring standards-based software, terminologies, and communications in all emergency management-related activities. While the focus of research and standards development is often on technical and semantic issues, interoperability is a multidimensional problem, with institutional, political, social, and legal ramifications. Indeed, the 9/11 Commission in its final report (National Commission on Terrorist Attacks Upon the United States, 2004, p. 418) concluded:

> Recommendation: The President should lead the government-wide effort to bring the major national security institutions into the information revolution. He should coordinate the resolution of the legal, policy, and technical issues across agencies to create a "trusted information network."

The National Spatial Data Infrastructure provides an excellent template for the development and sharing of geospatial data. It provides standards for the documentation of data resources (geospatial metadata standard); a network and procedures for posting, discovering, and accessing geospatial data (Geospatial One-Stop Portal and NSDI Clearinghouse Network); and base layers of framework data (National Map). While much progress has been made in implementing the NSDI, many agencies and organizations are not yet managing their geospatial data resources effectively or participating fully in the NSDI. Data description through metadata is often insufficient to support effective discovery, and conflicts exist between the metadata requirements of NSDI and other programs such as the National Pipeline Mapping System (NPMS) and the National Incident Management System (NIMS) Integration Center (NIC). A new effort to develop the necessary policies and guidelines for the support of emergency management, led by DHS but within the framework of the NSDI, would strengthen the efforts of both DHS and the NSDI and bring about more consistency. The recent work of DHS to develop the geospatial data model in conjunction with the FGDC is an example of how this could work and should be extended even further. Such a new initiative will likely require strong backing, in the form of a directive or even legislation, if it is to be effective. The Emergency Management Accreditation Program (EMAP), which supports the continued development of standards for emergency management, could be of great help in developing the needed standards, and the National Emergency Managers Association (NEMA) and the International Association of Emergency Managers (IAEM) also would be critical in their development, but even more so in helping to ensure that standards are adopted and disseminated to the emergency management community.

> **RECOMMENDATION 3: A new effort should be established, within the framework of the NSDI and its governance structure and led by DHS, to develop policies and guidelines that address the sharing of geospatial data in support of all phases of emergency management. These policies and guidelines should define the conditions under which each type of data should be shared, the roles and responsibilities of each participating organization, data quality requirements, and the interoperability requirements that should be implemented to facilitate sharing.**

4.3 GEOSPATIAL DATA SECURITY

The security of the data that are gathered for emergency management must be examined from a variety of perspectives. To begin with, there is the need to determine the actual risk to the nation should these data fall into the hands of terrorists or others with harmful intentions. A report[20] by the RAND National Defense Research Institute entitled "Mapping the

---

[20]*http://www.rand.org/pubs/monographs/MG142/index.html.*

Risks: Assessing the Homeland Security Implications of Publicly Available Geospatial Information" (Baker et al., 2004) found that fewer than 1 percent of federal data sets are unique and 94 percent of the data sets would not be useful to terrorists.

While this study clearly demonstrates the limited risk of making such data widely available, it has not overcome the real fear exhibited by some agencies about open access to their data, particularly in law enforcement, defense, and local government. Whether or not the perception is justified, it is real. Data that are not secure may be subject to intentional tampering as well as inadvertent corruption. Thus, many agencies believe that data shared for emergency management must be held securely, and if security cannot be guaranteed, many data custodians will be unwilling to provide their data. The committee heard from a panelist from New York State that efforts to collect geospatial data for homeland security purposes were more likely to succeed if security measures were guaranteed to the data custodian. If data are shared and security is subsequently breached, the resulting publicity could damage the agency's long-term effectiveness and erode its political support.

On the other hand, emergency planners at all levels must be aware that needless restrictions on access to geospatial data can lead to skepticism, if not open hostility, among local officials, the media, and the public. When restricted access reflects understandable agreements with private-sector data holders, the public needs to be informed about the reasons for these limitations and to be assured that key data will be available to those who need them should an emergency occur. Similarly, an informed public needs to appreciate restrictions on the release of unique, clearly sensitive, publicly held data involving critical infrastructure or hazardous materials. However, while narrow, sensible restrictions can inspire confidence and cooperation, overly broad restrictions on access, particularly when equivalent data are readily available through other channels, are unnecessary, as discussed in the Rand report cited above. Equally unwarranted are limitations on publicly collected geospatial data, which should be available to emergency responders and the public, and should include accurate metadata with a conscientious assessment of usability or strong, clear caveats regarding appropriate use.

The issue of security should be addressed whenever agreements are made for the sharing of data. Protection of confidentiality is one major source of concern in such agreements, because data about individuals and property that are useful during emergencies may be perceived as invasions of privacy at other times, when they might be used by criminals, for example. Appendix B presents a possible model for a confidentiality agreement that could be incorporated into negotiations over data sharing where appropriate.

THE CHALLENGE

It is important to note that some data elements used for emergency management may contain highly sensitive information (see Sidebar 4.1, for example), while other data elements may already be in the public domain. There is a need to be able to extract elements useful to emergency management without compromising security or criminal investigations. A system of security levels could be established, with appropriate rules governing access at each level, and applied to entire data sets, specific features in data sets, or specific attributes of those features as appropriate.

Once agreements are in place or guidelines have been established about the security of data that are shared for use in emergency response, methods must be implemented to carry out secure access to the data effectively. One model mentioned earlier is the National Pipeline Mapping System, which acts as a secure source of pipeline information. Another potential technique would be to utilize the role that One-Call centers play in managing underground critical infrastructure data from a variety of data sources. As mentioned earlier in this chapter, One-Call centers act as secure repositories of local utility information, which can be called upon to mark the locations of underground utilities before excavation work is done. Since these types of organizations have already established relationships with utility data owners and receive facility updates at scheduled intervals, they possess the means to potentially serve as a "one-stop" source for facility data. Such an approach would alleviate the need for data owners to provide their data to multiple agencies and would leverage existing processes that are already in place in the majority of regions throughout the United States. Although participation in a One-Call program is currently voluntary, the ability to participate in a data-sharing program to support national disaster recovery initiatives may serve as incentive to participate. In a recent example of an effort to address this issue, as mentioned earlier, the U.S. Geological Survey (USGS) in partnership with the Department of Homeland Security and the National Geospatial-Intelligence Agency (NGA) has implemented a central "GIS for the Gulf" database within the framework of the Geospatial One-Stop. When a federal emergency is declared, a user name and password will be distributed to all government agencies within the affected area, allowing them to access an extensive collection of geospatial data. Mechanisms also exist within this framework for agencies to contribute their own data, which will be available to others during emergencies under the same constraints.

Conclusion

Data security is a key element of any data-sharing effort in support of emergency management, and a mechanism that will protect and reassure the suppliers of data is therefore a core requirement. Guidelines defining appropriate levels of security for various kinds of data needed for emergency response have to be established and implemented. Again, the emergency management professional organizations such as NEMA and IAEM could be instrumental in helping support the development and adoption of guidelines.

> **RECOMMENDATION 4: DHS should lead, within the framework of the NSDI, the development of a nationally coordinated set of security requirements for data to be shared for emergency preparedness and response. All organizations should implement these guidelines for all data shared in support of emergency management and should use them where necessary to restrict access to appropriately authorized personnel. In concert with these efforts, the leveraging of existing organizations that could potentially serve as a "clearinghouse" for critical infrastructure data should be explored.**

4.4 OVERHEAD IMAGING

Disasters raise immediate questions about geographic extent or footprint, and about the intensity of impact within the footprint. Emergency responders at all levels need to know not only the areas affected and the severity of damage but also the locations of any people who might require timely rescue or immediate evacuation. When a disaster damages critical parts of the telecommunications or other infrastructures or when calls originating within and outside the area overload circuits, severely injured people might not be able to summon help. Also, because a disaster with a wide footprint readily overwhelms local first responders, agencies outside the region need to know the condition of the transport system, including places where rescue helicopters and other aircraft might safely land. Aerial surveillance of comparatively small sites is also beneficial, especially when wreckage or complex terrain thwarts line-of-sight observation from the ground. Existing geospatial data might describe the road network and pinpoint special-needs populations, but these are baseline data prior to the event, and effective emergency response requires a broad, up-to-date depiction of the devastation. However valuable, isolated reports from victims and rescue teams will not initially provide as complete a picture as images from an aircraft or a remote-sensing satellite with a high-resolution sensor.

Aerial images can expedite disaster response and recovery if they meet three requirements: (1) a strategically positioned platform collecting imagery at the right place and time, (2) a competently revealing imaging system (with sufficient geographic detail), and (3) skilled interpreters.

Because the timing of the disaster and the type of damage affect all three considerations, there are no one-size-fits-all solutions. For instance, a disastrous seismic event occurring in early morning a few hundred miles from an airport with reconnaissance aircraft and knowledgeable image specialists allows for timely assessment with photographic or imaging systems relying on reflected solar radiation. By contrast, overhead imaging is likely to be thwarted by an earthquake occurring late on a winter evening, with darkness approaching and a long night ahead. Many imaging satellites pass over only near local noon, at intervals that may be as long as two weeks. Delayed imaging is also inevitable when a severe, slow-moving tropical storm makes low-altitude flying dangerous and creates heavy cloud cover beneath comparatively safe, high-altitude flight paths. Even so, timely aerial coverage after the storm dies out or moves on is highly valuable to officials orchestrating response and recovery.

Four imaging platforms that are potentially useful during or shortly after an event are fixed-wing aircraft, remote-sensing satellites, helicopters, and unmanned aerial vehicles (UAVs). Examination of their diverse applications and limitations reveals significant complementarity. All four platforms are developing rapidly, particularly in those aspects that can potentially benefit emergency management, so they are discussed in some detail from that perspective in the following paragraphs.

The fixed-wing airplane is the most common platform for aerial imaging and mapping. Whether propeller driven or jet powered, conventional aircraft can reach a disaster scene within an hour or two from a base several hundred miles away and provide generally thorough coverage of a scene several hundred square miles in extent. Of particular interest are technological improvements that enable some level of rectification of the imagery (processing of the imagery so that the location of objects in the image is more nearly the same as their actual location on the ground) while the aircraft is in flight and telemetering of the imagery (rectified or not) from the plane to a receiving station on the ground, processes that can potentially reduce the time between image acquisition and delivery to responders by 24 to 48 hours. If turbulence is minimal, low- or medium-altitude aerial imagery might be acquired for multiple, slightly overlapping flight lines. When winds or unstable air make flying hazardous or preclude accurate imaging, aircraft with high-resolution photographic systems or multispectral scanners can capture suitable images from higher, safer altitudes. Cloudy skies are another obstacle, but radar-based imaging systems designed to penetrate cloud cover may also be useful at night depending on the ground cover and building density. When atmospheric conditions preclude conventional approaches and the need is urgent, use of high-altitude aircraft and satellite platforms could be crucial despite the inevitable loss of resolution.

Emergency response can also benefit from the improved resolution of imaging systems authorized for and available from commercial remote-sensing satellites in low-Earth orbit (altitude of roughly 400 to 800 kilometers). Commercial remote-sensing firms offer panchromatic, color infrared, true-color, and multispectral satellite imagery with resolutions of 1 meter (3.3 feet) or better. Off-nadir viewing, with the scanner directed sideways rather than straight down, has substantially reduced revisit times, which are significantly shorter than the 18-day cycle for the comparatively coarse imagery collected by Landsat-1, the pioneering civilian remote-sensing satellite launched in 1972, and may be as little as three days. Further improvements are likely, with private-sector imaging firms planning to launch "next-generation" satellites before the end of the decade that will collect imagery with even greater geographic detail, allowing objects as small as half a meter or less to be detected. As the level of detail improves and the number of commercial satellites increases, the remote-sensing industry has the potential to become an increasingly significant source of geospatial data for emergency managers.

By contrast, helicopters afford a much longer "dwell time" than either remote-sensing satellites, which must race along their orbital paths at thousands of kilometers per hour, or fixed-wing aircraft, which cover large areas systematically with carefully configured parallel flight lines but cannot hover over a specific site. Although helicopters can be equipped with aerial cameras and other imaging systems, photogrammetric firms prefer fixed-wing aircraft, which are more cost-efficient for mapping wide areas. Municipal and state governments that own helicopters typically use them only for rescue and law enforcement because commercial photogrammetric surveying is appreciably less expensive for routine mapping. Moreover, police agencies have been reluctant to conduct systematic aerial imaging since 2001, when the Supreme Court, in a strongly worded decision, warned against unconstitutional warrantless searches (Kyllo v. United States). While it is unlikely that helicopters equipped for medical evacuation could be diverted to imaging, other state and municipal helicopters might usefully be equipped with video imaging systems, which could be linked to an emergency operations center. In Los Angeles and other large municipalities where television broadcasters use helicopters for traffic reporting and news coverage, the private sector could be an important partner in real-time overhead imaging.

UAVs, also called drones, can be useful as well, especially during a radiological emergency, when low-altitude surveillance might imperil human pilots. UAVs appropriate for real-time overhead surveillance vary in size and payload. At one extreme are larger versions of the comparatively inexpensive, remote-controlled model aircraft used by hobbyists. Equipped with a small, ultralightweight video camera, a model airplane could record the scene below on magnetic media. Time aloft is limited, and the operator would have to keep the craft in view to orchestrate a safe landing or trigger a retrievable drop. A UAV with a slightly larger payload might carry more fuel for a longer range or a battery and transmitter for real-time video transmission. GPS could improve navigation and control, and a gyro stabilizer could allow more accurate imaging (see, for example, Oshman and Isakow, 1999). The military, which recognizes the importance of "over-the-hill" surveillance, has been experimenting with mini-UAVs for more than a decade.[21] By contrast, the Air Force's sophisticated high-altitude endurance class Global Hawk, which can attain an altitude of 65,000 feet and keep a payload of 1,960 pounds aloft for 42 hours, can be operated from a base hundreds of miles away.[22] Less expensive but no less relevant to emergency response imaging is the Air Force's medium-altitude endurance class Predator, which can fly above 40,000 feet and sustain 29 hours of flight with a 700-pound payload. UAVs equipped for communication with a satellite can operate hundreds of miles from their base. A smaller, less sophisticated military system is the joint tactical class Hunter, designed for real-time aerial surveillance with a range of up to 200 kilometers, a maximum altitude of 15,000 feet, and 12 hours of flight with a 200-pound payload. Deployment of UAVs at military bases around the country raises the possibility of timely imaging support with flexible UAV platforms and trained interpreters.

Efficient use of overhead imagery during an emergency will depend on the availability of trained personnel at the emergency operations center itself or at a remote site in direct contact with the center. Although identification of obvious obstacles such as a fallen bridge or blocked highway requires little expertise, trained interpreters are needed to recognize seriously weakened structures or signs of trapped occupants, both of which might require timely inspection by responders in the field. Communications are especially important because field personnel, especially those who know the area well, will have knowledge of significant benefit to experienced interpreters at a remote site. Because careful before-and-after comparison is a key strategy in image interpretation, baseline imagery acquired before the event is absolutely essential for change detection as well as for identification of trouble spots on which real-time video surveillance by helicopter or mini-UAV should be focused. Overhead imagery is also valuable for developing response plans, planning and carrying out training exercises, and planning mitigation efforts.

---

[21]For a description of the five-pound Dragon Eye used by the Marine Corps for over-the-hill surveillance, see *http://www.globalsecurity.org/intell/systems/dragon-eye.htm*.
[22]*http://www.fas.org/irp/program/collect/global_hawk.htm*.

There are several issues related to the use of overhead imagery for emergency management. First, there is a critical need for rapid data acquisition during emergency situations, and these temporal requirements often cannot be met, especially for data gathered using remote sensing. Issues of availability and level of geographic detail will continue to preclude the use of remote-sensing imagery in the immediate response phase of the disaster. However, the use of remote-sensing products for the preparedness and mitigation phases, when speed is not as critical, is also important. Three types of barriers inhibit the fuller and more immediate access to overhead imaging systems needed for emergency response and recovery. Contractual barriers exist where private-sector service providers are reluctant to commit staff and resources without guaranteed compensation. Contingency contracts available to municipal and state governments and specifying imaging requirements, payment schedules, and various options and fees can ensure the prompt cooperation of private photogrammetric mapping firms. Statutory barriers include restrictions, real or perceived, on cooperation between different levels of government or between civilian and military agencies. Federal agencies, on the other hand, may have indefinite delivery-indefinite quantity (IDIQ) contracts in place prior to an event that can permit very rapid development of mission assignments for image acquisition. Regulatory barriers include Federal Aviation Administration (FAA) restrictions on UAVs, which could interfere with other air traffic. Demonstrations of the reliability of UAVs and procedures for monitoring their use and licensing operators might overcome the FAA's understandable concerns, especially in times of emergency when special rules are needed for civilian aircraft (aware of the potential value of UAVs, the FAA is currently developing new policies, procedures, and approval processes for their regulation[23]). Another general issue brought to the committee's attention has been the lack of a lead federal agency to coordinate the procurement of imagery during response. While this has led to duplication of effort or confusion among the various private industry vendors contracted to acquire imagery, these problems currently are being addressed by the Federal Emergency Management Agency (FEMA) and DHS with USGS and NGA as the lead agencies.

Conclusion

Based on the experiences of committee members and workshop participants, the committee believes that overcoming barriers to the rapid acquisition of high-resolution imagery, particularly through contingency

---

[23]*http://www.faa.gov/news/news_story.cfm?type=fact_sheet&year=2005&date=092005.*

contracts, is crucial to its timely and effective use in disaster response. Remotely sensed imagery should be available for all phases of disaster management, and its acquisition should be better coordinated among federal agencies and enabled through the development of IDIQ contracts and through partnerships with federal agencies having such contracts. Advantage should be taken of advances currently being made in remote-sensing technology using the full range of fixed-wing, helicopter, UAV, and satellite platforms.

> **RECOMMENDATION 5: Standing contracts and other procurement mechanisms should be put in place at local, regional, and national levels by the responsible agencies to permit state and local emergency managers to acquire overhead imagery and other types of event-related geospatial data rapidly during disasters.**

4.5 COMMUNICATION OF REPORTS TO AND FROM THE FIELD

As noted and documented in previous sections of this report, data and information are critical to the responders and emergency managers dealing with a disaster. The committee heard that information may be known by first responders on the front line but not known by managers in the command post or the emergency operations center (EOC) or by other agencies. Cultural differences between various groups of responders, such as police and fire organizations, can also inhibit communication of information. This information may include such things as knowledge of road closures and inundated areas, specific information about damage to infrastructure, or the location of disaster victims trapped in homes or other buildings. Inability to communicate this information between responders, and between responders and managers, can delay critical action and add unnecessarily to loss of life, personal injury, and property damage.

Flow of information to management groups at higher levels within the responding agencies and the sharing of data with FEMA and other agencies are essential components of the response. The National Response Plan's Emergency Support Function (ESF) 5, discussed in Section 3.3.3, and the Homeland Security Operations Center (HSOC) serve the overall disaster information function, supporting planning and decision making at both the field or regional and the headquarters levels with information from all sources. All data of interest to those outside the agency collecting them should be provided to ESF 5 and the HSOC. Similarly, ESF 5 and the HSOC should be providing necessary data to other responding agencies (e.g., road outage information should be collected by municipalities or the state police and provided to ESF and HSOC, who would redistribute these data to all responding agencies for whom this is important.) Imagery acquired during disasters is provided to FEMA headquarters, the Joint Field Offices (JFOs), and the state EOCs, as well as to the USGS and the U.S. Army Corps of Engineers, which post the imagery for viewing and downloading over the Internet.

A major difficulty in the past has been the inability to move information from the field to the EOC or JFO, and then to the headquarters of the responding agencies and to FEMA. The technical components of these problems hinge on both computer security and the use of different data standards and formats. Data standardization, at least for those data used during events, will greatly facilitate the transmission of critical information. The EDXL standard described earlier in Section 4.2 will be very helpful in this regard. The committee also heard about specific problems resulting from the use of agency firewalls. For example, the FEMA firewall has impeded sharing of data during some incidents, which forced the FEMA GIS team to set up a separate network outside the firewall to share the data.

Difficulties also exist with the use of the imagery provided. While most disaster-related missions require high-resolution data, responders in the JFO, EOCs, and Emergency Response and Recovery Offices (ERROs) are typically unable to work with the huge files involved. To maximize their usefulness these data must be either compressed so that they can be pulled across the network at the field offices, written onto physical media such as CDs or fire wire hard drives and transported physically, or converted into hard-copy maps with reduced resolution. Implementation of a web-services architecture using NSDI practices would help alleviate this problem by enabling the transfer of only those portions of an image file that are actually needed by the user. Metadata, including indexing of imagery and other useful documentation, are especially helpful when data are to be transferred.

Perhaps the single greatest difficulty during disasters is the movement of files (large and small) between agencies working in the National Response Coordination Center (NRCC), the Regional Response Coordination Center (RRCC), and the JFOs when responders from agencies other than FEMA are attempting to communicate with their own agencies and attempting to connect to their home agency networks, and when telephone and Internet communication may be impossible. This issue is of paramount concern to computer security specialists.

Attempts to reduce these difficulties range from the marginally acceptable provision of analog phone lines at FEMA facilities, which let responders use modems to connect to their home facilities, to the use of high-speed digital subscriber lines (DSLs). While the second solution is significantly faster, DSL transmission still requires the use of two computers, one at the origin and one at the destination, and requires that both they and the network be operational and secured. The security problem may be solved in at least two ways. The first is to give disaster workers from another facility the same responsibility for protecting the network that they have at their own agency. If this strategy cannot satisfy computer security specialists, the gov.net concept (Sewell, 2002), whereby subnets on each of the agencies involved in the response become part of an intranet, might provide secure access but only to essential parts of each agency's network.

The culture of not sharing data and information, which has already been addressed in Section 4.2, is less tractable but of critical importance to reports to and from the field. Because of strong resistance, the issue will have to be dealt with through directives from higher levels in the organizations involved. The assistance necessary to help affected populations can best be delivered when all responders are able to contribute their information in timely fashion and when procedures are in place to see that the information is redistributed to those that need it and available for those who might need it.

An additional difficulty has been the communication (or its absence) between the federal and non-federal partners in disaster responses. Many states and municipalities complain of multiple data calls from federal agencies for nearly the same information. Coordinating data calls through one federal agency such as FEMA or USGS (who is putting liaisons in every state), and coordinating their geospatial needs through a single point of contact such as the state GIS coordinator, would help streamline operations. This as well as various cultural differences from different experience bases can best be dealt with through exercises that result in the development of partnerships prior to the onset of a disaster and also through the use of the incident command system mandated as part of the National Response Plan (NRP). Because the NRP is new, it may take some time (and many exercises) before these problems are resolved.

Conclusion

For a variety of reasons ranging from damage in the disaster area, to firewall issues, to other inter- and intrainstitutional problems and conflicts, data access in the field is often compromised. Plans must be in place for the provision of broad-band Internet and intranet services to emergency responders or for data transmission through other means such as physical transport of CDs or other digital media. Likewise, firewall issues that prevent access to essential data by members of multiple agencies located at disaster sites (e.g., a JFO, the FEMA National Response Coordination Center, FEMA's Regional Response Coordination Centers, or other locations not run by the responding personnel) must be resolved.

ESF 5 and the HSOC must continue to address the role of the collectors and providers of data and information to the responding community. This would be greatly facilitated by the use of standardized databases by all emergency responders, but also requires identification of procedures and deadlines for the provision of data and information to FEMA and DHS, as well as policies and procedures for pushing data and information to the responders and higher headquarters of all agencies. This will require procedures for making data available to the larger response community as well. More broadly, the committee finds that information flow between entities participating in disasters must be improved, particularly between responders in the field, field offices, and emergency operations centers.

This and earlier sections of the report have drawn attention to a host of problems that currently impede communication to and from the field. While many forms of action might help to address these problems, in the committee's view the best strategy would be to invest in intensive preparedness exercises, in which all aspects of communication can be tested.

RECOMMENDATION 6: Interpersonal, institutional, technical, and procedural communications problems that currently inhibit communication between first responders in the field and emergency operations centers, emergency management agency headquarters, and other coordinating centers should be addressed through intensive preparedness exercises by groups involved in all aspects of disaster management. Such exercises should be tailored to focus on clear objectives with respect to the use of geospatial data and assets. They should involve decision-making representatives from all levels of government, as well as other relevant organizations and institutions, and should be coordinated nationally so that common problems can be identified. They should be realistic in their complexity and should allow participants to work carefully through the geospatial challenges posed by disasters, including the difficulty of specifying requirements, the difficulty of communicating in a context of compromised infrastructure, and the difficulty of overcoming logistical obstacles.

4.6 BACKUP, REDUNDANCY, AND ARCHIVING

Ideally, all data could be accessed through a distributed network from local sources. This would guarantee that all responders are working off the latest version of the data. However, numerous experiences have shown that data can be lost, servers and networks can fail, and power and communication systems can go down. In order to guarantee effectiveness

THE CHALLENGE   117

during a disaster, multiple methods of accessing data have to be available and tested regularly.

Standardized methods are needed to ensure appropriate data backup and recovery. Of particular significance is the geographic dispersion of the backed-up data so that copies survive the event and can be used in response and recovery. On September 11, 2001, the New York City Office of Emergency Management lost both its primary and backup data because both were stored close to the World Trade Center (see Section 2.1.1). The committee also learned that backup copies of data for the New Orleans area were stored in close proximity to each other, and as a result, during Hurricane Katrina both the original and the backup copies of some data were destroyed.

As noted in Chapter 3, the data, tools, and procedures used during an event are rarely if ever archived. A number of individuals and organizations told the committee of the need to establish procedures to archive event-created data on a daily basis. This, in turn, would allow researchers at a later time to measure the effectiveness of the geospatial information and to determine methods to improve its contribution to the overall emergency response effort in future events. Since it is likely that some of the data used by responders will be proprietary, confidential, or subject to privacy laws, archiving plans will have to balance the desire for openness with security (see Section 4.3). Archiving plans will also have to include effective strategies for data management.

Conclusion

Backup and archival plans should exist for all geospatial data, tools, and procedures developed as part of disaster response and recovery, in order to ensure the security of essential resources through geographic dispersal, and to provide extremely useful knowledge for improving response to future events. Responsibility for this function should be stated in Emergency Support Function 5 of the National Response Plan. Since FEMA is currently given the responsibility in ESF 5 for coordinating GIS support, the archive and backup function should be stated as part of FEMA's responsibilities in the JFOs.

RECOMMENDATION 7: DHS should revise Emergency Support Function 5 of the National Response Plan to include backup and archiving of geospatial data, tools, and procedures developed as part of disaster response and recovery. It should assign responsibility for archiving and backup in the JFOs during an incident to the Federal Emergency Management Agency, with an appropriate level of funding provided to perform this function.

4.7 TOOLS FOR DATA EXPLOITATION

A variety of geospatial tools exist that can meet a wide range of emergency management needs. The following material describes these tools, how they are used, and issues that inhibit their successful exploitation. Since another recent NRC report has an extensive discussion of technologies and methods for disaster research that includes geospatial data, research, and technology (NRC, 2006), this report focuses on impediments to the take-up of existing tools and the need to adapt them better to the conditions of emergency management.

Visualization tools provide the opportunity to visualize features of the pre- and post-event world individually or simultaneously. These features may include hazards (fault zones, potential hurricane landfall areas, and flood inundation areas) and risks (potential hurricane damage zones based on projected wind speed and its impact on population, building types, and critical infrastructure). These may be stand-alone tools for operation on a single system or server-based tools designed for Internet or intranet use, allowing an operator in one location to view data stored on a server at a remote location. It is also possible to perform data-mining operations on the data through the display-tool interface to determine critical relationships between hazards, disaster events, and the most appropriate actions.

Analysis tools include a wide range of models performing a hierarchy of functions, from models indicating impact area and expected severity (shaking or wind speed), to those showing expected damage (combining shaking or wind speed with geology and construction type), to models to determine evacuation routes based upon the road network and traffic flow (see Figures 4.1, 4.2, and 4.3). They can include validated atmospheric models such as those used by the National Hurricane Center; atmospheric plume models used by the national laboratories; waterborne plume models used by the National Oceanic and Atmospheric Administration (NOAA); earthquake and wind damage models used by FEMA, the Department of Defense, and others; hydraulic and hydrologic models used by the U.S. Army Corps of Engineers; or empirical models used by the Corps to estimate debris volume from disasters. They also include high-interaction computer graphics, three-dimensional virtual reality media, and other visualization techniques that promote realistic simulations of disastrous events and planned responses.

Decision support systems assist emergency and other managers in making the best decisions based upon conditions as they are known at a particular point in time. These systems are often a combination of display capabilities, one or more models, and visualization and data analysis functions. With high-resolution digital elevation models, hydrology and hy-

FIGURE 4.1 Example of modeling of ground shaking in the northwestern United States. Courtesy of the Washington State Emergency Management Division and the Federal Emergency Management Agency.

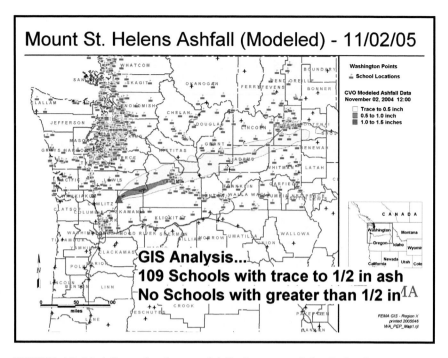

FIGURE 4.2 Modeling of volcano ashfall. Courtesy of the Federal Emergency Management Agency.

draulic models, elevations for the living floors of dwellings, and economic damage models based on the depth of water above the elevation of the living floor, it is possible to model the consequences of the operation of water control structures during high-flow events. Both upstream and downstream inundation with different release rates from the water control structure can be displayed, and analysis of the economic damage that occurs for any release rate option can be calculated. While economic damage is not the only factor involved in release rate decisions, this is an example of how such systems can help the decision-making process. As part of a multiagency effort, the U.S. Army Corps of Engineers is participating in the development of a decision support system that will help communities and regulatory agencies evaluate the consequences of different potential land-use decisions in a river subbasin on communities located downstream.[24]

---

[24]The High Resolution Land Conversion, Hydrology, and Water Quality Modeling of Large Watersheds is a joint project of the U.S. Army Corps of Engineers' Construction Engineering Research Laboratory, Coastal and Hydraulics Laboratory, and the Corps' Rock Island District, in conjunction with local communities. A draft report was due in spring 2006.

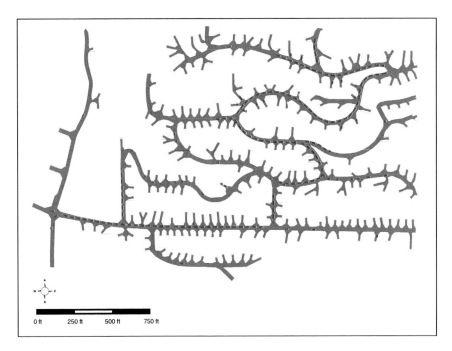

FIGURE 4.3 Screen shot of an evacuation simulation for an area of Santa Barbara, California, called Mission Canyon, where red indicates individual vehicles. Microscale traffic simulation suggests that congestion will be great enough that if everyone drives their cars out during an evacuation, a fire may overtake the neighborhood while cars are still attempting to leave. Simulations such as this allow investigators to evaluate a number of policy options, including new traffic controls and reductions in the number of cars used for evacuation. Courtesy of Richard Church, University of California, Santa Barbara.

Geospatial models and tools are currently being used successfully in numerous areas. One excellent example is the National Interagency Fire Center in Boise, Idaho, that controls forest fire fighting using a range of GIS tools, has a clear organizational structure, and sends real-time information about topography, vegetation, and weather forecasts—and the resulting fire predictions—to field units rapidly using handheld systems. Another successful system presented to the committee was by the Province of Alberta, in which GIS and geospatial data are an integral part of its full emergency management operations system. These successful systems often concerned single disciplines or small jurisdictions where it was possible to avoid the kinds of interagency issues addressed earlier in this report. Yet despite these successes, there are still many impediments to the better use of geospatial tools in emergency management, as described below.

First, all of these possibilities are easier to realize when the necessary data are readily available and easily accessible. Ideally, they will take advantage of interoperability between the systems used by the different entities involved with a disaster event. At their best, network-enabled approaches will allow (with sufficient security) all emergency responders to access data sets and analytical products that are located on servers managed by other responding entities. However, present implementations of geospatial tools are largely, if not entirely, the work of single agencies and not easily distributed. Although such solutions are workable, they are inefficient in the larger context.

Moreover, during emergency response, data may well be incomplete and of poor quality, and first responders may be working under very difficult circumstances with limited technical resources. It is very difficult and indeed unlikely that response personnel will take on the task of learning about new tools in this type of situation. Therefore, training on tools must take place so that people working in an emergency situation feel completely comfortable with their use (see further discussion on this topic in Section 4.8). Furthermore, in the report *Making the Nation Safer* (NRC, 2002, p.162), a section on information management and decision support tools makes the following comment:

> In a chaotic disaster area, a large volume of voice and data traffic will be transmitted and received on handheld radios, phones, digital devices, and portable computers. Nevertheless, useful information is likely to be scarce and of limited value. Thus, research is needed on "decision-support" tools that assist the crisis manager in making the most of this incomplete information.

In some cases, the modeling capabilities exist even though the needed input data are not readily available. For example, loss estimation tools exist; however the underlying nature of the default data on the building inventory, infrastructure, and economic structure of places means that only very generalized estimations can occur, normally at a regional scale. The damage is expressed as the probability of the building being in one of four damage states: slight, moderate, extensive, and total, with a range of generalized damage functions (expressed as repair cost or replacement cost) assigned to each damage class. The loss estimations work best when applied to a class of buildings (e.g., residential), rather than to individual buildings. Moreover, specific loss estimations at the city or county level require supplemental data on building stock and infrastructure, localized data not normally included. Finally, the loss estimations represent "direct" damage, not "indirect" losses such as lost wages, loss of business earnings, or loss of building use, in the overall loss estimation.

As one emergency management professional told the committee, a serious impediment to better use of geospatial data and tools for disaster management is that "uninformed, overwhelmed public officials get sold expensive systems they don't need and don't know how to use." Several testified to the need for a "common denominator" set of tools with designs based on user requirements. Moreover, such tools must be simple, easy to use, and tailored to what users really need (for example, functions to assist navigation through the application, functions for basic query and measurement of location, and tools for the management of saved files). At the same time, users often fail to take advantage of capabilities because they are unfamiliar with them. Typically, users only encounter geospatial data and tools during emergencies, so they do not know what is available or how to make use of it. Also, it can be hard to get users to adopt new technology, especially in the midst of an event when novel approaches feel like distractions rather than solutions to emergency responders. To address these challenges, users argue that geospatial data and tools should be integrated as a routine component of emergency planning, training, exercises, and routine incident operations, so that during major disasters they are readily available and easily incorporated.

Conclusion

Although numerous tools exist and may be very useful in the planning stages, they are not as effectively used during response because (1) the necessary data may be of poor quality or not available during response, or (2) the tools have not yet been fully integrated into regular response activities. The committee concludes that efforts should be made to more effectively integrate the use of geospatial tools into all phases of emergency management, as proposed in Recommendation 1. Additional research is needed on how geospatial data and tools can be used for decision support in the special conditions that prevail during emergency response.

> **RECOMMENDATION 8: The National Science Foundation and federal agencies with responsibility for funding research on emergency management should support the adaptation, development, and improvement of geospatial tools for the specific conditions and requirements of all phases of emergency management.**

4.8 EDUCATION, TRAINING, AND ACCESSING HUMAN RESOURCES

Presentations to the committee provided ample evidence of the nonuse and underutilization of geospatial data and tools, and previous sections of this report have focused on many of the causes cited by the individuals and agencies that provided testimony. This section focuses on one of the more important and endemic causes: the lack of appropriate education and training in geospatial data and tools among emergency management personnel and a similar lack of education and training in emergency management among geospatial professionals. These deficiencies exist at all levels, from the bottom of agencies to the top, and must be addressed by programs that raise awareness among leaders as effectively as among their staff. It is important, moreover, to recognize that education in the many issues surrounding the use of geospatial data and tools in emergency management that are identified in the report, as well as in underlying principles of geospatial data, such as the correct interpretation of maps, their time-limited nature, and knowledge of their inherent uncertainties, is at least as important as training in the technology itself.

While academic emergency management programs at both the bachelor's and the master's levels are growing in the United States, the committee heard that the emphasis given to geospatial data and tools varies widely. Geospatial data and tools are not always considered essential from a curricular standpoint and must compete for space in the curriculum with many other subject areas. As an earlier NRC study concluded, "The very people who could leverage this information [geospatial data and tools] most effectively, such as policy makers and emergency response teams, often cannot find it or use it because they are not specialists in geospatial information technology" (NRC, 2003, p. 3). Only by requiring that those in emergency management programs take classes in geospatial data and tools (with a primary focus on emergency management applications), and by including modules in other emergency management classes showing how geospatial data and tools can be used in all phases of emergency management, will it become clear to future generations of emergency managers that geospatial data and tools have significant contributions to make. Furthermore, because geospatial information is often taken at face value, and since the response community is getting greater access to multiple sources of geospatial data, it is critical to ensure that the underlying assumptions, data quality, and uncertainty are conveyed properly. Geospatial data and tools are often so complex that even geospatial professionals sometimes lack the training for proper interpretation of results. Access to technical specialists is key to the proper use and improvement of such mapping products. For example, it is critical to understand the time sensitivity of technical information (e.g., decaying radiation dose from deposited material, relative changes in indoor and outdoor exposure during a hazardous airborne release).

In a similar way, geospatial data and tools must be a component of the in-service training offered to the current generation of emergency management professionals. While such content is currently included in some of the training offered by FEMA at its Emergency Management Institute in Emmitsburg, Maryland, and by the U.S. Army Corps of Engineers in training provided to members of its emergency management planning and response teams, there is a need to include such content in all relevant training programs nationwide. First responders need relatively rudimentary training in geospatial capabilities: they have to be able to communicate what conditions they encounter and what they need to know to fulfill their mission assignments. Incident command- and management-level personnel (e.g., plans and operations section chiefs) need a more sophisticated understanding of geospatial capabilities. Both users and GIS personnel are extremely busy, stressed, and sometimes emotionally volatile during emergency response. This is not the best time to assess needs or to learn new material. Data sources and tools should be presented to emergency management personnel before an incident in a training situation so that they know what will be most useful. The concepts, as well as the end products, have to be documented.

It is similarly important that geospatial professionals are acquainted with the emergency management process and that geospatial products are designed to be useful to emergency managers. Many geospatial professionals who become involved in emergency response are not routinely associated with the emergency management community in their normal roles, however. It is essential therefore that training be provided as part of emergency preparedness or, at worst, that it be part of the orientation process when geospatial professionals or volunteers join response and recovery teams. Not only should this training explain the emergency management organization in which the geospatial team will operate, but it should also provide a window into the pressures and time constraints under which personnel will be expected to perform. Emergency responders do not always know how best to articulate their information and imagery needs to geospatial professionals. Sometimes they are unsure of what they need because they do not know what is possible. It is important for technical personnel to spend time helping responders frame their questions. Sometimes, it helps to produce a variety of products and let responders identify those that are most helpful. Working with emergency management personnel in this way can help provide geospatial professionals with feedback on how well geospatial data and tools address the needs of responders, especially in terms of ease of use, interpretational and supporting information, and documentation.

The lack of sufficient personnel trained in all aspects of the application of geospatial data and tools to emergency management is a problem at all levels of government, from local to federal, although specific needs vary from agency to agency. In particular, trained personnel with knowledge of imagery products should be included as essential members of the emergency management team. The past two hurricane seasons have shown that the ability to provide these personnel has already been stretched very close to the breaking point. Annual or semiannual exercises that provide the opportunity for involved agencies to meet, exercise, and discuss potential geospatially related successes and pitfalls in the event of an actual disaster can both raise awareness of the importance of an adequate supply of personnel and provide essential experience to those involved.

Disasters, by definition, overwhelm the ability of local emergency managers to respond sufficiently, and recent disasters have demonstrated the importance of being able to augment local human resources with professionals and volunteers drawn from both neighboring and remote areas. One of the common problems reported to the committee was the lack of a preestablished team of geospatial professionals to support emergency response within a significant number of emergency management organizations. As a result, when a catastrophe occurs, a significant amount of effort and time is wasted to locate geospatial professionals, bring them into the emergency management organization, and provide them with resources to accomplish their mission. By the time they become available, many of the opportunities to apply technologies to solve problems have passed.

As the committee heard repeatedly and as noted in Section 3.2.2, FEMA has only a limited number of permanent geospatial professionals and must rely on reservists to respond to events, which almost inevitably delays deployment. If federal geospatial professionals arrive on scene after state and local staffs have already begun work, it can be difficult to integrate and coordinate the various efforts. There is clearly a need for FEMA to expand the effective size of its permanent staff of geospatial professionals, perhaps through dual use, and to develop strategies that will lead to their more rapid deployment.

In addition to having a preestablished team of geospatial professionals, having a mechanism in place to locate additional geospatial professionals to respond to a disaster is essential. Subsequent to the attacks on the World Trade Center, the members of GISMO (Geographic Information Systems and Mapping Operations), the New York City GIS user group, used their contact lists to assist in assembling a team of volunteers. Having this type of information can be invaluable in a disaster that is geographically large and runs over an extended period of time (see Sidebar 4.3). A national system to facilitate access to additional geospatial professionals from a range of related fields could be organized by the Department of Homeland Security, perhaps in partnership with universities as nodes of expertise, and used during disasters to locate and assemble

> **Sidebar 4.3**
> **"Geocoding" Used to Locate Katrina Survivors—Street Addresses Not Very Useful After Hurricane Hit**
> *By Marsha Walton, CNN*
>
> Police, firefighters, and Coast Guard crews may be the first to come to mind when naming the lifesavers during disasters such as Hurricane Katrina. It might be time to add geographers to that list. In the sometimes desperate hours following Katrina's landfall, experts in geographic information services—GIS—helped search and rescue crews reach more than 75 stranded survivors in Mississippi. One of their most valuable tools was a process called "geocoding," the conversion of street addresses into GPS coordinates. With streets flooded, street signs missing, and rescue crews unfamiliar with the Gulf Coast area, street addresses were not very useful.
>
> "They would get phone calls, or the Coast Guard would come in with addresses in their hands and say, 'I need a latitude and longitude for this address.' So the GIS professionals would do a geocoding, give it to the Coast Guard who got on helicopters and saved lives," said Shoreh Elhami, director of GISCorps.
>
> Elhami, co-founder of GISCorps, said that since 2004, the organization's volunteers have responded to disasters such as the Asian tsunami and Hurricane Katrina, as well as efforts to provide humanitarian relief, sustainable development, economic development, health, and education in all parts of the world. The Corps had 20 volunteers on the ground in Mississippi less than 48 hours after Katrina's landfall. GISCorps is part of URISA, the Urban and Regional Information Systems Association. Elhami said more than 900 qualified volunteers have GIS experience, and range from city and state government officials to academics to people in private industry.
>
> Volunteer Beth McMillan, a field geologist and professor at the University of Arkansas in Little Rock, worked in Pearl River County, Mississippi, a couple of weeks after the storm. "A couple of days after the hurricane hit, I felt so down, and wondered what I could do. I could give a little bit of money, but that doesn't seem very satisfying. To be able to have a skill that can be used is much more empowering, it doesn't make you feel so helpless," said McMillan, back in Little Rock. . . .
>
> Volunteers are never sure of the conditions they might face when deployed to disaster sites or developing countries. Assignments usually last between two weeks and two months. McMillan said her many experiences "roughing it" as a field geologist helped her deal with the living conditions in Mississippi. "They said be prepared for really hot weather, and bring a sleeping bag," she said. "I slept in an empty U.S. Department of Agriculture building on a cot, with probably several hundred other people. But it did have power, bathrooms, and showers, so conditions were not as bad as they could have been," she said. . . .
>
> SOURCE: Excerpted from Walton (2005).

teams. Such a system could also promote appropriate training, by establishing and implementing the minimal qualifications needed to be listed.

RECOMMENDATION 9: Academic institutions offering emergency management curricula should increase the emphasis given to geospatial data and tools in their programs. Geospatial professionals who are likely to be involved in emergency response should receive increased training in emergency management business processes and practices.

RECOMMENDATION 10: The Federal Emergency Management Agency should expand its team of permanent geospatial professionals, and develop strategies that will lead to their more rapid deployment both in response to events and in advance of events when specific and reliable warnings are given.

RECOMMENDATION 11: The Department of Homeland Security should establish and maintain a secure list of appropriately qualified geospatial professionals who can support emergency response during disasters.

4.9 FUNDING ISSUES

Along with the lack of an effective governance process (see Section 4.1), funding is usually identified as a major barrier to effective use of data in preparing for and responding to disaster events. For many organizations, particularly those in states and localities that are comparatively resource poor, there is inadequate funding to build even a basic geospatial capability. For others, funding is lacking for ongoing programs to maintain and update existing geospatial data, for the servers and support services needed to ensure effective access to and use of these data for converting data formats to meet a standard, and for creating the metadata needed to make data accessible through the NSDI. Finally, others lack the capability of participating in coordination activities due to shortage of personnel and funds. At the state level, geospatial preparedness is often not seen as sufficiently important to qualify as a target for funds that flow from federal homeland security programs. Several points are evident: different locations have different sets of needs and requirements across the country; not all of the perceived funding problems can be fixed by the infusion of more money; and funding for geospatial investments has often not been seen as a high priority.

A National Academy of Public Administration (NAPA) study, conducted in the late 1990s and titled *Geographic Information for the 21st Cen-*

tury: Building a Strategy for the Nation (NAPA, 1998), identified geospatial information and technology as key components of substantial elements of the U.S. economy. The report cited some of the major sectors of the economy that are impacted by geographic information and stated that geographic information plays a role in about one-half of the economic activities of the United States (NAPA, 1998, p. 11). Although not focused on budget or funding issues, the report also stated that competing priorities at the time, such as the year 2000 computer problem, created the reality that in the absence of additional major funding, only part of the highest-priority efforts would be implemented and that fulfillment of the stated goals for the NSDI was many years away (NAPA, 1998, p. 66-67).

It is not known with any accuracy how much money is spent by the many units of government for geospatial activities. In part this is due to differences in the programming and budgeting systems that exist among the nation's levels of government, along with the fact that much of the nation's resource of geospatial data and tools is acquired and used as part of a mission program, not as a specifically identified activity. However, the Government Accountability Office (GAO) estimated that billions of dollars are spent each year by federal agencies and their partners on geospatial data, services, technology, and expertise (GAO, 2004), and a global figure for 2000 of $12 billion to $20 billion was given by Longley et al. (2001, p. 360). With this amount of money being appropriated annually to sustain the existing—but in many ways inadequate—resource of geospatial data and tools, it is essential that ways in which current funding could be used more effectively be found, in addition to calling for new funding.

As noted in Section 1.1, the past two decades have seen dramatic increases in the use of geospatial data and tools in many aspects of human activity. Data needed for emergency management are often collected, managed, and disseminated for other purposes, particularly in the case of the framework and foundation data defined in Section 1.3.2. Collection of variables particularly important for emergency management might be piggybacked on existing data collection activities at minimal additional cost, and similar economies might be found in the costs of data dissemination.

Many changes will be needed in existing practices if geospatial preparedness is to be funded more adequately. While this has proven difficult, some components of government have made significant progress by

- Adopting a clear strategic direction that lays out future objectives;
- Initiating changes early within the organization in order to address personnel, structural, or other adjustments that affect employee performance;

• Planning a funding bridge to enable transition from current business processes to new ways of conducting business; and
• Establishing customer expectations that help to drive the needed changes but are also realistic about the pace at which changes can be made.

The committee heard in testimony that the most critical gaps in current funding appear to be

• A lack of funds that can be used in shared arrangements to leverage the funding resources of multiple organizations;
• A lack of funds for coordination activities among multiple organizations; and
• The lack of a long-term base of funding to sustain geospatial data collection, maintenance, and dissemination over time.

Several previous efforts have been explored to address these needs, and it is useful to examine them as potential guidance in any renewed attempts to address financing issues, whether through new mechanisms, legislation for grant programs, or increases in agency appropriations.

The FGDC conducted a study in 2000 to explore alternative mechanisms for "Financing the NSDI."[25] The study report found that an opportunity existed to provide a national capacity through public-private partnerships that could underwrite information technology investments for geospatial data and tools, and could provide the capital financing that local, regional, industry, and interest group consortia need to form and grow. The report recommended ways in which these consortia could pool and align intergovernmental and public-private investments in geospatial data acquisition and maintenance; decision support applications; and supporting hardware, software, and integration services. Financial mechanisms such as government-backed bonding authority for use by local governments, revolving loan programs, and other debt structures were suggested for use in a range of capital planning strategies. Financing would be dependent on the use of consensus standards for interoperability from recognized standards development organizations and other NSDI design elements as underwriting criteria.

Other NRC reports have also discussed funding options for the NSDI and related issues (NRC, 1993, 1994), and several legislative proposals have identified increased funding as a need. As one example among many, in 2003 a committee formed within the Spatial Technologies Indus-

---

[25]*http://www.fgdc.gov/publications/fgdcpubs.html.*

try Association (STIA) identified the need for increased funding and a government-wide legislative base for establishing and maintaining geospatial preparedness for homeland security, national defense, electronic government, and other purposes. The proposal was presented in testimony to the House Committee on Government Reform by STIA President Fred Corle.[26] One of the key elements identified was a major grant program to assist non-federal levels of government to build and maintain the NSDI and to achieve geospatial preparedness. Key elements of the grant program were that it was to provide matching funds as an incentive for geospatial preparedness and would require participants to adhere to standards for interoperable access, sharing, and use as part of the development and implementation of the NSDI.

To make it easier for organizations to find grant programs that they can utilize to obtain funding for various activities, the federal government has taken steps recently to identify grant opportunities through its electronic government initiatives. The FGDC has a grant program called the NSDI Cooperative Agreements Program to assist the geospatial data community through funding and other resources in implementing the components of the NSDI.[27] This program is open to federal, state, local, and tribal governments, and to academic, commercial, and nonprofit organizations and provides small seed grants to initiate sustainable ongoing NSDI implementations. This program could be used for geospatial preparedness activities. The Department of Homeland Security has a number of grant programs for emergency management in which geospatial activities could be included as part of the applicant's proposal. However, there is no comprehensive grant program that would provide funds for coordinated actions across the nation to better organize, manage, share, and use the geospatial data and technology that exist now and are being acquired for emergency management and other important public purpose and business reasons.

Conclusion

The committee concludes that the funding available to achieve geospatial preparedness for disasters is not sufficient to meet the need. Adequate resources must be made available to support existing mandates and new initiatives that integrate geospatial resources into all phases of emergency management and facilitate the acquisition and sharing of geospatial data for emergency management. In particular, resources such

---

[26]http://reform.house.gov/UploadedFiles/Corle%20Testimony1.pdf.
[27]http://www.fgdc.gov/grants/.

as grants need to be made available at the state and local levels where many of the emergency management activities occur, but where resources may be lacking to adequately support the development of the geospatial capabilities needed.

> **RECOMMENDATION 12:** To address the current shortfall in funding for geospatial preparedness, especially at the state and local levels, the committee recommends: (1) DHS should expand and focus a specifically designated component of its grant programs to promote geospatial preparedness through development, acquisition, sharing, and use of standard-based geospatial information and technology; (2) states should include geospatial preparedness in their planning for homeland security; and (3) DHS, working with OMB, should identify and request additional appropriations and identify areas where state, local, and federal funding can be better aligned to increase the nation's level of geospatial preparedness.

5

Guidelines for Geospatial Preparedness

5.1 INTRODUCTION

The preceding chapters of this report have identified numerous issues, and Chapter 4 made certain specific recommendations. The discussion to this point has been general and has included issues that affect all levels of government, the private sector, nongovernmental organizations, and the general public. It seemed to the committee that it would be useful at this stage of the report to assemble a set of guidelines, based on presentations to the committee and the content of previous chapters, that could be used by agencies seeking to improve their levels of geospatial preparedness. The chapter has been assembled by members of the committee with particular experience in emergency operations and presents a series of potential solutions to the issues encountered in the previous chapters. Appendix C presents a checklist developed from those issues and solutions that is designed to assist members of the emergency management community in analyzing how well they have integrated geospatial data and tools into their emergency management processes and how well prepared they are to take advantage of them during a disaster. Its goal is to help all levels of government to understand better their deficiencies in this arena and to suggest actions they might implement to be better prepared for future disasters. While the lists here and in the appendix represent a compilation of issues and topics addressed by the committee, made available to it through submissions and testimony, and compiled from the relevant literature, they are nevertheless incomplete. The committee fully expects that organizations and individuals may find it necessary or appropriate to make additions to the lists.

5.2 CRITICAL ELEMENTS IN SUCCESSFUL PLANNING AND RESPONSE

To ensure a successful planning and response effort for emergencies, there are a set of critical elements that are important at all response levels from the local to the national. Testimony to the committee identified a number of geospatial elements that were especially relevant. These are identified and explained below.

5.2.1 Integration

- *Develop standard, written operating procedures that integrate the use of geospatial information across all phases of emergency management at the municipal, state, and federal government levels.* Where necessary, modify existing procedures to incorporate the use of geospatial information into the workflow and decision-making cycle of emergency managers at all levels as well as first responders.
- *Develop a geospatial team location at the emergency operations center.* Provide a dedicated workspace, data, hardware, software, and infrastructure to support a geospatial team.
- *Establish a close working relationship between the state geographic information systems (GIS) coordinator and the state emergency management staff.* Hold regular meetings between the state GIS coordinator and his or her emergency management counterparts to determine gaps between resources and needs. Develop an action plan to bridge those needs.
- *Establish relationships and coordinate activities.* Establish working relationships with adjoining jurisdictions and between state and federal governments to share data and products prior to an event occurring. Colleges and universities, as well as national labs, also may have centers of geospatial expertise that could provide support in an emergency. If useful geospatial information is to be provided to the emergency response community in an incident, then all geospatial communities must be included in the process early on. At the local level, relationships must be established with geospatial professionals at the county and state levels to ensure that a coordinated approach is developed for the sharing of data prior to an event and for the distribution of products during an event. In addition, a methodology for obtaining regular inventories of useful local, county, state, and federal data must be established through the data custodians.

All of these relationships should ideally be built first at the local community level, then at the county level, then the state level, and finally the federal level. If done properly, they will establish a robust network that will enable the data custodians and the GIS coordinators to add value to the process and to assist all levels of the emergency response community in the development of a dynamic infrastructure to support their needs.

In particular, federal government organizations intending to provide geospatial services for an area need to coordinate their activities better with state, county, and municipal GIS coordinators as well as their emergency management counterparts to save time and resources and to eliminate duplication. In most cases, this coordination can be best done by working through the state GIS coordinator.

- *Develop agreements.* Develop agreements between geospatial professional teams at the municipal, state, and federal levels to predetermine the data and products to be used, generated, and shared during a disaster. Additionally, determine the roles on which each level will concentrate in order to avoid duplication of effort when a large event occurs.

- *Obtain around-the-clock contact information for GIS coordinators and their emergency management counterparts along with their respective backups in the state.* Make this information available to the emergency management community at all levels in the state. In addition, provide around-the-clock contact information for the emergency management counterpart (and his or her backup) to the GIS coordinator in the state. Develop similar contact information for the GIS coordinators and their emergency management counterparts in each county and metropolitan government in the state.

- *Develop a secure web site with around-the-clock contact information for GIS coordinators, their emergency management counterparts, and their respective backups in each state.* This web site should be publicized and made available to federal agencies that respond to emergencies as well as to emergency responders and GIS coordinators across the country. It would allow emergency responders and GIS coordinators to communicate more easily before, during, and after an incident to determine information on human resources, data, and equipment across their state.

5.2.2 Human Services

- *Establish a geospatial team to support emergency response.* A geospatial team should be created for an emergency response organization. This team needs to be organized with roles established for each team member. These roles should include an overall manager, a liaison (to coordinate activities and determine needs of responders), and highly skilled technical staff. All participants should be quick and able to work under extreme pressure.

- *Establish a geospatial away team.* One of the lessons learned from the events of September 11, 2001, was the value of having geospatial professionals close to the site to support operations. Where possible, having geospatial professionals that can be used in a team to complement efforts at or near the incident site can be extremely valuable both to provide responders with detailed geospatial information and to retrieve incident-related data from the site. These teams should be equipped at a minimum with data, hardware, and software. In some larger urban areas and states, these teams need permanent vehicles to support this role.
- *Develop an up-to-date inventory of geospatial professionals in each state along with their areas of expertise that can be called upon to respond to an emergency.* This inventory should be shared between the state GIS coordinators and their emergency management counterparts and should include around-the-clock contact information for each person (see Recommendation 11).
- *Establish a team to do modeling or establish contact with national teams having this responsibility.* Establish or make contact with teams of experts with scientific expertise to develop and run models for plume analysis, hurricane surges, flooding, and so forth. Include on the team both geospatial experts able to use the available software and scientists with backgrounds in a variety of areas able to ensure proper input into the models and reasonable results.

5.2.3 Training

- *Establish a training program for all levels of emergency response across the country that details the decision-making support available from geospatial tools.* Conduct training on the capabilities for decision support that geospatial information can provide for emergency responders at all levels. Include information on the differing types of information applicable to each level of responder. Modify existing emergency management training to incorporate the use of geospatial information into the workflow and decision-making cycle of emergency managers at all levels and of first responders. To be productive and avoid the pathologies of many poorly executed programs, these exercises should have clear objectives related to geospatial operations and should be narrowly tailored to focus on these objectives.
- *Establish a training program for members of the geospatial response teams within each state to train them in emergency management procedures as well as the data and tools available for use in an incident.* In addition, establish a training program for other potential geospatial responders who may be called upon during a large event, detailing the emergency management organizational structure and Standard Operating Procedures. Both groups should be trained on the existing data, data-gathering methodologies and

technologies, the software, the equipment, and product delivery mechanisms to be used during an emergency.

- *Conduct regular scenario-based training exercises that incorporate the use of geospatial responders and geospatial information in supporting emergency response operations at all levels within the organization.* Use these exercises to familiarize emergency responders with the capabilities of geospatial information and geospatial professionals with the specific needs and timing for delivery of this information. In particular, these exercises should stress to geospatial responders both the need to understand the data clearly and the need for rapid delivery of geospatial maps and other products to emergency responders. Include the geospatial professional team manager and liaison in the exercise meetings and briefings to allow him or her to understand better the contexts in which the geospatial products are being used in the decision-making process.

Results of these exercises should be made available to those not able to participate, giving them the opportunity to learn from such experiences. Scenario-based exercises relying on the use of geospatial information should occur at least on an annual basis between municipal, state, and federal responders (for an example of such exercises, see the New York State web site),[1] and might leverage the National Exercise Program currently coordinated by the Department of Homeland Security (DHS) Office of Grants and Training.

5.2.4 Data Access

- *Relationships.* Gaining access to varied kinds of data required to respond to an emergency requires a concerted effort. The emergency management community and GIS coordinators need to work closely with the data custodians early in the process to understand their concerns and be able to accommodate them while meeting their emergency management needs.

Access to data should, where practical, extend from the local government up to county, state, and federal governments. State GIS coordinators should be involved in the process of developing a system to access those data easily and quickly for emergency response activities.

5.2.5 Data Quality

- *Establish a team to identify and gather data required to meet state emergency management needs.* Working with the emergency managers, this team should identify and locate the most appropriate data available to meet

---

[1]*http://www.nysgis.state.ny.us/outreach/training/gen_scenario_resp.htm.*

emergency response needs. After reviewing the data, the team needs to determine the quality and usability of the data. Where necessary, it should identify data requiring improvement or data not currently available needing development. The team should coordinate these activities with the state GIS coordinator and the state emergency management agency. A joint plan should be developed that prioritizes the work on the data required to meet those needs.

Often data will require geographic adjustments that can be made relatively easily by overlaying the data on recent digital orthoimagery. In other cases, there will be deficiencies in the data's attributes. These issues should be addressed by working closely with the data custodians to correct or complete missing information. In all cases, the improved data should be returned to the data custodians for their use and to encourage improved data updates down the line (see Section 5.2.7).

5.2.6 Data Gathering

- *Establish agreements with local governments and utilities and other private-sector sources.* Work closely with local governments and utilities and other private-sector data sources to establish good relationships to share data. Where required, establish legal agreements to obtain access to and use of data required from local governments for emergency response use. Where possible, these agreements should be worded to allow use of those data by multiple organizations at various levels of government for training, preparation for, and recovery from emergency events. In addition, the time frame for updates should be established and agreements reached for obtaining those updates. If possible, on-line access to the latest data should also be arranged.
- *Develop a GIS-based system to track the distribution of emergency equipment and supplies.* This system should be established to enable emergency managers to route and track the location, quantity, and type of items stockpiled in and distributed to an area. It will greatly improve distribution of supplies and equipment to those in need, enhance their ability to recover reusable equipment (emergency generators and pumps), and provide information that can be analyzed to determine future needs (see Sidebar 5.1).
- *Develop and deploy a system for electronic field collection of data.* Obtaining timely, accurate, on-site data depicting the extent of an incident or the nature of its impact is critical to assist emergency managers in planning their response to an event. One method for accomplishing this is by using Global Positioning System (GPS) enabled, handheld computers with wireless communication systems. If these computers are programmed for emergency events and have drop-down menus to describe the character-

Sidebar 5.1
Homeland Security Secretary Announces Tracking System for Emergency Equipment and Supplies[a]

Speaking on February 13, 2006, to the National Emergency Management Association Mid-Year Conference, Secretary Michael Chertoff addressed some of the problems that became apparent in the response to Hurricane Katrina, including the lack of effective tracking of emergency supplies:

> Despite this remarkable effort, FEMA's [the Federal Emergency Management Agency's] logistics systems were not up to the task of handling a truly catastrophic event. The reality is, FEMA lacks technology and information management systems to effectively track shipments and manage inventories. FEMA relies on other government agencies, like the Department of Transportation, who often serve as agents of FEMA, and contract through their extensive network of private sector entities to provide support and move most of the necessary commodities. But the fact of the matter is, if FEMA is going to take responsibility for moving goods and services, it can't do it by remote control. It has to have the ability directly to impact the way in which we monitor and supervise and are able to effect in real time the movement of those supplies. Therefore, DHS must have some of the same skill sets that 21st century companies in the private sector have to routinely track, monitor and dispatch commodities where they are needed.
>
> Our first step for strengthening FEMA will be to create a 21st century logistics management system that will require the establishment of a logistics supply chain, working with other federal agencies in the private sector. What that means in the very short-run—because we're not going to get all this done immediately—is that we have to put agreements into place before the need arises again to ensure a network of relief products, supplies and transportation support that can be tracked and managed. In other words, we are going to insist this year, as we go into contracting, that we are going to have as a capability with anybody who is carrying our goods and services real-time visibility to where those deliveries are, when they're going to arrive, and, if necessary, the ability to redirect those, if the emergency so requires.

---

[a]http://www.dhs.gov/xnews/speeches/speech_0268.shtm.

istics encountered, these data can be gathered relatively easily and forwarded quickly to the emergency operations center.

- *Store the inventoried data within each state's emergency operations center in at least one geographically distant location.* Because communication networks and power systems are often impacted by large events, data stored in the emergency operations center should also be backed up in at least one geographically distant location. Updates to all stored data (including backups) should be scheduled on a regular basis. Where possible, this should include metadata and a data catalog. Last, the alternate locations of these data should be provided on a secure web site for use by other authorized users, if required.
- *Securely store copies of the same data on physical media capable of being brought to incident sites for response and recovery efforts.* Provide the locations of these media on the secure web site.
- *Establish an agreement with a national site to store and provide secure on-line access to a copy of these data.* Once again, redundancy of data storage is being stressed to accommodate the many problems with data access that have occurred in past disasters. This national site should be an "industrial-strength" site and should specialize in data storage. One site that should be considered for such a role is the U.S. Geological Survey's Center for Earth Resources Observation and Science (EROS) in Sioux Falls, South Dakota.
- *Establish regional or statewide emergency contracts to provide imagery of an incident site within 24 or at worst 48 hours of an event.* One of the first types of data needed after an event is aerial imagery. Contracts have to be in place to allow rapid access by all levels of government as soon as they have provided the appropriate level of funding to cover the costs.
- *Establish agreements and standard operating procedures to acquire digital images via state, county, municipal, or private-sector helicopters, et cetera, within one hour of an event and covering impacted sites.* Digital pictures taken from a helicopter can provide valuable insight into the extent of an incident for emergency managers. While they may have limited use as geospatial information, they provide a very quick perspective of the event. During the World Trade Center recovery efforts, they were used on a regular basis along with aerial, thermal, and LIDAR (light detection and ranging) imagery. Establishing agreements for helicopter support prior to an event with local police or state agencies (or the private sector) can be extremely valuable.
- *Develop an up-to-date inventory of municipal, county, and state data available for emergency response in each state (including metadata).* This inventory should note the best data available in all categories, utilizing the resources of the National Spatial Data Infrastructure (NSDI) as much as possible. It should include contact information for individuals familiar

with how the data were developed. This inventory should be posted to a secure web site based on NSDI architecture that is accessible by emergency managers. Where possible, the inventoried data and metadata should be made available on-line as well.

- *Distribute these inventories to municipal, county, state, and federal responders and the state GIS coordinators.* In addition, provide them with access to this information on the secure web site. Finally, develop a methodology to keep these inventories up-to-date on an annual basis, at a minimum.

5.2.7 Data Improvement

- *Establish a joint federal, state, and municipal program to fund the development and annual update of critical geospatial data.* The data developed or improved under such a program must be coordinated through the state GIS coordinators and meet the needs of local governments. Such a program should be flexible in providing funding and resources that supplement existing local and state government efforts. Where data do not exist, efforts should be made to provide resources or funding to the state GIS coordinator to allow their creation and to ensure that the data created meet local government needs.

5.2.8 Information Delivery

- *Deliver data rapidly.* One of the keys to making geospatial information a valuable resource to the emergency management community is to be able to deliver it when it is needed. When an event occurs, emergency responders want to react quickly to limit its effects. Therefore, they are looking for significant information on that event within the first hour of it happening. Traditionally, a number of reasons cause delay in providing geospatial information, including gathering the team, locating the incident, and creating maps to portray its location, extent, and impact. If this information is to be included and to be of value in early decision making, the geospatial team must deliver certain standard products to the emergency managers rapidly when they are establishing their initial strategies.
- *Understand the needs.* The geospatial team must understand the needs of emergency responders and learn to match the appropriate products with those needs. If a responder needs to know the schools (and contact numbers) in an area to be evacuated, a simple list may be more useful and certainly quicker than printing out a beautifully composed map of those school locations. Close coordination with the emergency managers and knowledge of their workflow and information needs by the geospatial

team manager are essential in understanding their requirements and choosing the appropriate output.

- *Develop standard work products.* Develop protocols that deliver standard work products to emergency managers quickly through the use of templates and other simple programs. Adoption of a standard map layout should be part of this process. The goal of the team should be production of maps within the first hour after the event occurs or shortly thereafter.

In New York State, an application has also been built to allow emergency managers to query geospatial information and obtain maps and reports through a web application without any significant technical expertise. Making this capability directly available to emergency responders is another way to deliver this information faster. A similar initiative has been taken in South Carolina.[2]

- *Enhance models depicting the impact of disasters hitting the community.* Models for incidents that can potentially occur in the community should be adapted to the community's location and context. In coastal areas, models can be run to determine the worst-case storm surge for a particular hurricane category. This, in turn, can provide an idea of the population that may have to be evacuated and the logistical requirements for evacuating and sheltering that population. Similar models can simulate each of the hurricane categories that could hit the area. In other areas of the country, models could be developed to determine the extent of flooding from a river based on the predicted height of the water in that river. This predetermined information can then be archived for use during a real incident at a later time. Development of these models should be coordinated at the state level to avoid redundancy and should be made available to authorized geospatial professionals at all levels of government.

- *Keep it simple.* Because most emergency managers and responders are not fully familiar with geospatial data and tools and because they are under extreme pressure during an incident, they may lack the time to understand clearly what is being presented to them. This problem can be reduced if the team remembers to keep its products simple and uses graphics that make them easy to understand.

- *Anticipate needs.* While providing the standard products quickly is essential, it is important to be able to anticipate other needs specific to the type of event as it continues. One way to do this is by having the geospatial team liaison take time to meet with operational groups and

---

[2]http://www.cas.sc.edu/geog/hrl/scemdmain.htm.

task forces within the emergency operations center to determine their needs and inform the team during an event. At the same time, the liaison can offer these groups suggestions on how geospatial information might provide them with assistance.

- *Test delivery mechanisms.* The finest geospatial products in the world will be useless if they cannot be delivered in a timely manner during an emergency. Some delivery methods can be as simple as capturing an image and pasting it into a PowerPoint presentation. There are some that appear simple, such as printing out a map, but that can become agonizing if the plotter fails to respond properly or is too slow. Other mechanisms, such as posting data to a web site or moving it to a web-based services application can be limited by numerous technical issues. All of these methods should be tested on a regular basis to ensure their availability in a time of need.
- *Practice.* There is no substitute for the team practicing the delivery of these products. Once the team is satisfied that it can deliver the standard products quickly, then it needs to incorporate its workflow into scenario exercises. These can be done separately with the team and then as part of the emergency operations center workflow. If this practice is approached properly, team members will be better prepared to deal with the pressure of a real-life situation and able to react more effectively to it to meet emergency managers' needs. It is important to keep logs describing the methods used and examples of the maps and other products generated to support retrospective analysis of performance.

5.2.9 Hardware and Infrastructure

- *Establish a secure, nationwide methodology to access local data.* As an event grows from a local to a state and then to a national crisis, there is a need for more and more geospatial responders from different parts of the state and the country to provide support. However, this support cannot be given without access to the best data. A mechanism should be established to provide a consistent, secure nationwide methodology for responders across the country to obtain access to the best data easily and rapidly. The mechanism should accommodate multiple authorized users at different levels across the country. This mechanism must include appropriate redundancies and provide linkages to the appropriate data sources in each state. It must allow local, county, state, and federal governments to post and retrieve data without significant restrictions during an emergency. Whatever mechanism is chosen, it should be coordinated through the state, county, and local GIS coordinators. In addition, once established, this infrastructure should be integrated into training exercises for all levels of emergency response to ensure that this technology can be used effectively during a disaster.

- *Develop an up-to-date inventory of the relevant geospatial hardware and resources available for use in an emergency in each state.* This inventory should be developed in conjunction with the state GIS coordinator and should include resources from government, academic, and private-sector organizations. Around-the-clock contact information should be included for each resource. This inventory should also be posted to a secure web site for access by authorized emergency and geospatial responders.
- *Establish a backup satellite communications system.* Communication systems are often interrupted during disasters. Establishing a backup satellite communications system to transmit voice and geospatial data can be extremely useful to obtain reports quickly from the field and to transmit information to other locations around the state or the country.

6

Concluding Comments: Looking to the Future

While the emphasis throughout the committee's study and this report has been on geospatial data and tools, it is not surprising that there are similarities in many of its recommendations to the conclusions reached by a parallel and recently published study of information technology (IT) in disaster management, conducted by the National Research Council's Computer Science and Telecommunications Board (NRC, 2005). Writing about interoperability in communications during emergency response, that report concluded (NRC, 2005, p. 2):

> Most communications interoperability issues are not technical. Better human organization, willingness to cooperate, and a willingness of governments at higher levels to listen to those at local levels who really do the work and who are the actual responders are all critical factors in making better use of information technology for disaster management.

On the issue of training, the report concluded (NRC, 2005, p. 4):

> To be useful in a disaster, IT must be in routine use. In a crisis situation, people tend to fall back on what they are familiar with. Technology that is not included in planning, training, exercises, and standard operating procedures will not be used in an actual disaster.

The committee also recognized these issues, which were raised many times by participants in its meetings and workshop, and they underlie several of its recommendations, including those regarding the need for increased training and more effective exercises, for technology that is better adapted to the special circumstances of emergency management, and for more effective sharing of data within the framework of the National Spatial Data Infrastructure. In short, most of the problems associated with the use of geospatial data and tools are institutional and not technical, and the use of geospatial data and tools must be habitual rather than exceptional.

Geospatial data and tools have evolved over the past four decades into an extremely powerful and effective application of digital technology, and they are now widely deployed in many areas of human activity. However, emergency management provides a very different context for their application. First and foremost, its demands occur under enormous pressure of time—rather than weeks or months, data must be acquired and analyzed within minutes. Second, the users who rely on the products of geospatial tools are often poorly trained in their use and working under very difficult circumstances. Little room exists for error and little admiration for complex technologies that fail under pressure.

The central message of this report is that geospatial data and tools are useful and indeed essential in all phases of emergency management. Yet it is never easy to persuade authorities or the general public of the need for investment in information infrastructure when the primary concerns in the immediate aftermath of an event are clearly focused on food, shelter, and the saving of lives. The fact that information is essential to effective response—that "it all starts with a map"—is easily forgotten. The need for geospatial data and tools may be everywhere, but in a sense it is also nowhere in minds that are overwhelmed by the circumstances of a disaster.

The answer to this fundamental dilemma is clear, and the chapters and recommendations of this report will hopefully give it the exposure that it needs. Society must plan in advance for disasters and cannot afford to wait until the next one happens, as it inevitably will. Investment in infrastructure is an important part of preparedness, and the kind of infrastructure represented by geospatial data and tools is a very important part of that investment. The committee hopes that the experience of recent disasters, when emergency management agencies were so clearly overwhelmed by the sheer magnitude of the event, along with increasing awareness among the general public and decision makers of the potential usefulness of geospatial data and tools, will help to drive home the report's central message.

Another event of the magnitude of Hurricane Katrina will occur at some point in the future, and the hypothetical events described in Chapter 2 will almost certainly become real. When they do, and if the problems identified in this report and evident in the response to Hurricane Katrina are not addressed, similar patterns of breakdown will undoubtedly occur. However, with the kinds of preparedness outlined in this report, the events will occur in a very different world from that of 2005. Agencies will have planned the immediate and coordinated acquisition of data, using arrangements that are already in place and relying on technologies that are fully interoperable. Teams of geospatial professionals will be activated immediately, even prior to events if accurate warning exists, and will be on-site and operational within hours. Emergency response personnel will have practiced the use of geospatial data and tools under a range of scenarios and will be fully familiar with the kinds of problems they will encounter. Response will be better targeted and managed, and additional lives may be saved. The way to achieve this vision of preparedness is clear; society has only itself to blame if it is not realized.

References

Alexander, D. 2000. *Confronting Catastrophe*. Oxford, U.K.: Oxford University Press.

Baker, J. C., B. Lachman, D. Frelinger, K. M. O'Connell, A. C. Hou, M. S. Tseng, D. T. Orletsky, and C. Yost. 2004. *Mapping the Risks: Assessing the Homeland Security Implications of Publicly Available Geospatial Information*. Santa Monica, Calif.: RAND Corporation.

Bruzewicz, A. J. 2003. Remote Sensing Imagery for Emergency Management. Pp. 87-97 in S. L. Cutter, D. B. Richardson, and T. J. Wilbanks (eds.), *The Geographical Dimensions of Terrorism*. New York: Routledge.

Burton, I., R. W. Kates, and G. F. White. 1993. *The Environment as Hazard*, 2nd Edition. New York: Guilford Press

Clarke, K. C. 2003. *Getting Started with Geographic Information Systems*, 4th Edition. Upper Saddle River, N.J.: Prentice Hall.

Cutter, S. L., ed. 2001. *American Hazardscapes: The Regionalization of Hazards and Disasters*. Washington, D.C.: Joseph Henry Press.

Cutter, S. L. 2003. GI Science, Disasters, and Emergency Management. *Transactions in GIS* 7(4):439-445.

DeMers, M. N. 2005. *Fundamentals of Geographic Information Systems*, 3rd Edition. New York: Wiley.

DHS (Department of Homeland Security). 2005. *Challenges in FEMA's Map Modernization Program*. Washington, D.C.: Department of Homeland Security Office of the Inspector General, OIG-05-44.

Donahue, A. K., and P. G. Joyce. 2001. A Framework for Analyzing Emergency Management with an Application to Federal Budgeting. *Public Administration Review* 61(6):728-740.

ESRI (Environmental Systems Research Institute). 2001. GIS for Homeland Security: An ESRI White Paper. Redlands, Calif. Available at http://www.esri.com/library/whitepapers/pdfs/homeland_security_wp.pdf [accessed October 25, 2006].

FGDC (Federal Geographic Data Committee). 2001. Homeland Security and Geographic Information Systems. Washington, D.C.: Department of the Interior. Available at http://www.fgdc.gov/library/whitepapers-reports/white-papers/homeland-security-gis [accessed October 25, 2006].

Field, E. H., H. A. Seligson, N. Gupta, V. Gupta, T. H. Jordan, and K. W. Campbell. 2005. Loss Estimates for a Puente Hills Blind-Thrust Earthquake in Los Angeles, California. *Earthquake Spectra* 21(2):329-338.

Galloway, G. E. 2003. Emergency Preparedness and Response—Lessons Learned from 9/11. Pp. 27-34 in S. L. Cutter, D. B. Richardson, and T. J. Wilbanks (eds.), *The Geographical Dimensions of Terrorism*. New York: Routledge.

GAO (Government Accountability Office). 2004. Geospatial Information: Better Coordination Needed to Identify and Reduce Duplicative Investments, Washington, D.C. Available at http://www.gao.gov/new.items/d04703.pdf [accessed October 25, 2006].

Goodchild, M. F. 2003. Geospatial Data in Emergencies. Pp. 99-104 in S. L. Cutter, D. B. Richardson, and T. J. Wilbanks (eds.), *The Geographical Dimensions of Terrorism*. New York: Routledge.

Greene, R. H. 2002. *Confronting Catastrophe: A GIS Handbook*. Redlands, Calif.: ESRI Press.

Haddow, G. D., and J. A. Bullock. 2003. *Introduction to Emergency Management*. Boston, Mass.: Butterworth-Heinemann.

Kelmelis, J. A., L. Schwartz, C. Christian, M. Crawford, and D. King. 2006. Use of Geographic Information in Response to the Sumatra-Andaman Earthquake and Indian Ocean Tsunami of December 26, 2004. *Photogrammetric Engineering and Remote Sensing* 72(8):862-876. Available at http://www.asprs.org/publications/pers/2006journal/august/feature2.pdf [accessed October 25, 2006].

Longley, P. A., M. F. Goodchild, D. J. Maguire, and D. W. Rhind. 2001. *Geographic Information Systems and Science*, 1st Edition. New York: Wiley.

Longley, P. A., M. F. Goodchild, D. J. Maguire, and D. W. Rhind. 2005. *Geographic Information Systems and Science*, 2nd Edition. New York: Wiley.

MacFarlane, R. 2005. A Guide to GIS Applications in Integrated Emergency Management. London: Emergency Planning College, Cabinet Office. Available at http://www.ukresilience.info/publications/gis-guide_acro6.pdf [accessed October 25, 2006].

Mandia, S. 2005. The Long Island Express/The Great Hurricane of 1938. Available at http://www2.sunysuffolk.edu/mandias/38hurricane/index.html [accessed October 25, 2006].

NAPA (National Academy of Public Administration). 1998. *Geographic Information for the 21st Century: Building a Strategy for the Nation*. Washington, D.C.: NAPA.

National Commission on Terrorist Attacks Upon the United States. 2004. *The 9/11 Commission Report: Final Report of the National Commission on Terrorist Attacks upon the United States*. New York: W.W. Norton. Available at http://www.gpoaccess.gov/911/ [accessed October 25, 2006].

National Governors' Association. 1979. *1979 Emergency Preparedness Project: Final Report*. Washington, D.C.: National Governors' Association Office of State Services.

National Governors' Association. 2006. State Strategies for Using IT for an All-Hazards Approach to Homeland Security. NGA Center for Best Practices Issue Brief, July 13. Available at http://www.nga.org/Files/pdf/0607HOMELANDIT.PDF [accessed October 25, 2006].

NRC (National Research Council). 1993. *Toward a Coordinated Spatial Data Infrastructure for the Nation*. Washington, D.C.: National Academy Press.

NRC. 1994. *Promoting the National Spatial Data Infrastructure Through Partnerships*. Washington, D.C.: National Academy Press.

NRC. 2002. *Making the Nation Safer*. Washington, D.C.: National Academy Press.

NRC. 2003. *IT Roadmap to a Geospatial Future*. Washington, D.C.: The National Academies Press.

REFERENCES

NRC. 2005. *Summary of a Workshop on Using Information Technology to Enhance Disaster Management.* Washington, D.C.: The National Academies Press.

NRC. 2006. *Facing Hazards and Disasters: Understanding Human Dimensions.* Washington, D.C.: The National Academies Press.

Oshman, Y., and M. Isakow. 1999. Mini-UAV Attitude Estimation Using an Inertially Stabilized Payload. *IEEE Transactions on Aerospace and Electronic Systems* 35(4):1191-1203.

Sewell, C. 2002. One Network Under Gov. *Telephony* 242(1):30-34.

Stage, D., N. von Meyer, and R. Ader. 2005. *Parcel Data and Wildland Fire Management* (prepared for the FGDC Cadastral Data Subcommittee). Washington, D.C.: Federal Geographic Data Committee. Available at http://www.nationalcad.org/showdocs.asp?docid=149&navsrc=Report&navsrc2= [accessed October 25, 2006].

Thomas, D. S. K., S. L. Cutter, M. E. Hodgson, M. Gutekunst, and S. Jones. 2003. Use of Spatial Data and Geographic Technologies in Response to the September 11 Terrorist Attack on the World Trade Center. Pp. 147-162 in *Beyond September 11th: An Account of Post-disaster Research.* Special Publication 39. Boulder, Colo.: Natural Hazards Research and Applications Information Center, University of Colorado.

Walton, M. 2005. Geocoding Used to Locate Katrina Survivors: Street Addresses Not Very Useful After Hurricane Hit. Available at http://www.cnn.com/2005/TECH/11/10/gis.technology/index.html [accessed October 25, 2006].

Waugh, W. L., Jr. 1988. Current Policy and Implementation Issues in Disaster Preparedness. Pp. 111-125 in L. K. Comfort (ed.), *Managing Disaster: Strategies and Policy Perspectives.* Durham, N.C.: Duke University Press.

Waugh, W. L., Jr. 2000. *Living with Hazards, Dealing with Disasters: An Introduction to Emergency Management.* Armonk, N.Y.: M. E. Sharpe.

Worboys, M. F., and M. Duckham. 2004. *GIS: A Computing Perspective*, 2nd Edition. Boca Raton, Fla.: CRC Press.

Appendixes

Appendix A

List of Acronyms

ANSI	American National Standards Institute
API	application programmer interface
C-Coast	Coastal Cartographic Object Attribute Source Table
CAD	computer-assisted design
CAPWIN	Capital Wireless Integrated Network
CD	compact disc
CI/KR	critical infrastructure/key resource
CIPI	Critical Infrastructure Protection Initiative
COP	common operating picture
COTS	commercial off the shelf
DHS	Department of Homeland Security
DMA	Disaster Mitigation Act
DSL	digital subscriber line
EADRCC	Euro Atlantic Disaster Response Coordination Center (NATO)
EDXL	Emergency Distribution Exchange Language
EMAP	Emergency Management Accreditation Program
EMDC	Emergency Mapping and Data Center
EMS	emergency medical service
EPA	Environmental Protection Agency
EOC	emergency operations center

EROS	Earth Resources Observation and Science
ERRO	Emergency Response and Recovery Office
ESF	emergency support function
ESRI	Environmental Systems Research Institute
FAA	Federal Aviation Administration
FDNY	New York Fire Department
FEMA	Federal Emergency Management Agency
FFRDC	federally funded research and development center
FGDC	Federal Geographic Data Committee
FX	field exercise
GAO	Government Accountability Office
GDIN	Global Disaster Information Network
GECCo	Geospatially Enabling Community Collaboration
GIS	geographic information system
GISMO	Geographic Information Systems and Mapping Operations
GITA	Geospatial Information Technology Association
GMO	Geospatial Management Office
GOCO	government-owned, contractor-operated
GOS	Geospatial One-Stop
GPS	Global Positioning System
hazmat	hazardous material
HSIN	Homeland Security Information Network
HSOC	Homeland Security Operations Center
HSPD	Homeland Security Presidential Directive
IAEM	International Association of Emergency Managers
IC	incident commander
ICS	incident command system
IDIQ	indefinite delivery-indefinite quantity
IMAAC	Interagency Modeling and Atmospheric Assessment Center
ISO	International Organization for Standardization
IT	information technology
JFO	Joint Field Office
LIDAR	light detection and ranging
MAC	Mapping and Analysis Center
MSEL	master scenario events list

NADB	National Asset Database
NAPA	National Academy of Public Administration
NASA	National Aeronautics and Space Agency
NATO	North Atlantic Treaty Organisation
NEMA	National Emergency Managers Association
NGA	National Geospatial-Intelligence Agency
NGO	nongovernmental organization
NGS	National Geodetic Survey
NIC	NIMS Integration Center
NIEM	National Information Exchange Model
NIMS	National Incident Management System
NIPP	National Infrastructure Protection Plan
NOAA	National Oceanic and Atmospheric Administration
NOS	National Ocean Service
NPMS	National Pipeline Mapping System
NRC	National Research Council
NRCC	National Response Coordination Center
NRP	National Response Plan
NSDI	National Spatial Data Infrastructure
NSGIC	National States Geographic Information Council
OASIS	Organization for the Advancement of Structured Information Standards
OGC	Open Geospatial Consortium
OMB	Office of Management and Budget
ORNL	Oak Ridge National Laboratory
PCII	Protected Critical Infrastructure Information
PfP	Partnership for Peace
PIMS	PfP Information Management System
QA/QC	quality assurance/quality control
RRCC	Regional Response Coordination Center
RSDE	residential structures damage estimation
SDSS	spatial decision support system
SLOSH	Sea, Lake, and Overland Surges from Hurricanes
SOP	Standard Operating Procedure
STIA	Spatial Technologies Industry Association
TCL	Target Capabilities List
TTX	tabletop exercise

UAV	unmanned aerial vehicle
USDOT	U.S. Department of Transportation
USGS	U.S. Geological Survey
UTL	Universal Task List
VBMP	Virginia Base Mapping Program
XML	Extensible Markup Language

Appendix B

Sample Confidentiality Agreement

This Confidentiality Agreement ("Agreement") is effective as of the [Enter Day of Month] day of [Select Month] 2003 by and among

Name of Company Possessing the Data to Be Shared
 (hereinafter referred to as "Company")

and

[Enter Name of Company]
[Enter Type of Company]
Having its principal office at [Enter Address of Company]
(hereinafter referred to as "Contractor")

WITNESSETH:

WHEREAS, the Parties hold information and data that are proprietary to each Party respectively and desire to share certain confidential and proprietary information with each other in connection with [Enter Brief Description of Subject Matter] (hereinafter referred to as "Matter").

NOW, THEREFORE, the Parties agree as follows:

1. Definitions

    a. "Confidential Information" includes any and all information, including but not limited to all oral, written, graphical, and electronic information disclosed to the Party receiving the information. If Confidential Information is disclosed in written, recorded, graphical, electronic, or otherwise in a tangible form, it may be labeled as "proprietary" or "confidential" or with a similar legend denoting confidentiality, or it may otherwise be verbally designated as such.
    b. "Company" includes description of company, affiliates, and subsidiaries.
    c. "Party" or "Parties" refers to Company, Contractor A and Contractor B, individually and collectively.

2. The Parties agree not to use or disclose Confidential Information except for the purpose of the Matter. The Parties agree only to disclose the Confidential Information received from each other to the Parties' respective employees whose duties justify their need to know such Confidential Information. The Party disclosing Confidential Information shall ensure compliance by its employees with the terms and conditions of this Agreement.

3. Confidential Information is not information that

    a. Now is or becomes generally known to the public without fault of the Party or Parties receiving the information; or
    b. Is proven by written documentation to have been in the receiving Party's possession prior to its receipt from the disclosing Party; or
    c. Is received from an independent third party who is not under obligation of confidentiality.

4. To the extent the Party receiving Confidential Information is required by an order of a court of competent jurisdiction to reveal such information, the Party will promptly notify the Party that provided the Confidential Information in order to allow the Party to take necessary action including a protective order, as appropriate, and will cooperate with the disclosing Party in protecting the confidentiality of the Confidential Information in a lawful manner.

5. Disclosure of Confidential Information by any Party under this Agreement does not grant the receiving party any right or license to use the Confidential Information unless explicitly set forth

herein or in a letter of authorization from the disclosing Party and signed by an employee of that Party authorized to grant such authorization.

6. All Confidential Information, unless specified in writing, remains the property of the disclosing Party, and must be used by the receiving Party only for the purpose intended by the disclosing Party. Upon termination of this Agreement, all copies of written, recorded, graphical, electronic, or other tangible Confidential Information must be returned to the disclosing Party. The disclosing Party may in its sole discretion direct the receiving Party to destroy and certify in writing that it has destroyed the Confidential Information.
7. Confidential Information supplied is not to be reproduced in any form except as required to accomplish the intent of the Matter.
8. All Confidential Information must be retained by the receiving Party in a secure place with access limited to only such of the receiving Party's employees (or agents or subcontractors who have a non-disclosure obligation at least as restrictive as this Agreement) who need to know such information for the purposes of the Matter, and to such third parties as the disclosing Party has consented to by prior written approval. The receiving Party must provide the same care to avoid disclosure or unauthorized use of the Confidential Information as it provides to protect its own confidential and proprietary information.
9. Each Party warrants that it has the right to disclose the Confidential Information that it will disclose to the other Parties pursuant to this Agreement, and each Party agrees to indemnify and hold harmless the Parties from all claims by third parties relating specifically to the subject matter of this Agreement. Otherwise, no Party makes any representation or warranty, express or implied, with respect to any Confidential Information. No Party is liable for indirect, incidental, consequential, or punitive damages of any nature or kind resulting from or arising in any manner whatsoever in connection with this Agreement.
10. The Parties acknowledge that a receiving Party's unauthorized disclosure of Confidential Information may result in irreparable harm. The Parties, therefore, agree that in the event of a violation or threatened violation of this Agreement, without limiting the rights and remedies of each Party to seek damages, a temporary restraining order and/or an injunction to enjoin disclosure of Confidential Information may be sought against the Party who has breached or threatened to breach this Agreement and the

Party who has breached or threatened to breach this Agreement agrees not to raise the defense of an adequate remedy at law.
11. All media releases, public announcements, and demonstrations by any Party to this Agreement relating to the Matter, its subject matter, or the purpose of this Agreement must be approved in writing in advance signed by all Parties prior to the release, announcement, or demonstration.
12. No party shall assign any of its obligations under the Agreement without the prior written consent of the Company, which shall not be unreasonably withheld or delayed.

   a. Not withstanding the foregoing, the Company shall have the right to assign this Agreement to an entity as a result of merger, acquisition, reorganization, or sale of substantially all of the Company's assets.

13. The obligation to hold Confidential Information confidential is perpetual and shall survive this Agreement.
14. This Agreement represents the entire understanding between the Parties, and the terms and conditions of this Agreement supersede the terms of any prior agreements or understanding, express or implied, written or oral.
15. This Agreement may not be amended except in writing signed by all Parties.
16. The provisions of this Agreement are considered to be severable, and in the event that any provision is held to be invalid or unenforceable, the Parties intend that the remaining provisions will remain in full force and effect to the extent possible and in keeping with the intent of the Parties.
17. There are no additional party beneficiaries to this Agreement.
18. Failure by a Party to enforce or exercise any provision, right, or option contained in this Agreement will not be construed as a present or future waiver of such provision, right, or option.

Accepted by	Accepted by
[Enter Name of Company]	[Enter Name of Company Possessing the Data to Be Shared]
By:	By:
Name:	Name:
Title:	Title:
Date:	Date:

Appendix C

Preparedness Checklist

The attached checklist is designed to allow federal, state, county, and municipal governments to assess their use of geospatial information and tools to improve their first responder and emergency management capabilities. Ideally, the committee would like this checklist to be used by the emergency response communities in each municipality, county, and state working together with the state geographic information system (GIS) coordinator. This should also include federal government agencies that are called on to respond to emergencies, such as the Federal Emergency Management Agency (FEMA), the U.S. Army Corps of Engineers, and the Coast Guard.

**Measuring Emergency Management Preparedness:
A Checklist to Determine Your Ability to Use Geospatial Information in Emergency Management**

Integration

- ❏ Does your emergency operations center (EOC) have geospatial technology available?
- ❏ Do you have a permanent workspace or office for your geospatial team?
- ❏ Is the use of geospatial information integrated into your emergency management operations and used in emergencies?

- ❏ Do your written standard operating procedures include the use of geospatial information in your workflow and decision-making processes?
- ❏ Do you know the name of your state GIS coordinator?
- ❏ Do you have contact information for the state GIS coordinator and his or her backup?
- ❏ Does your state GIS coordinator know who his or her emergency management counterpart is in your organization?
- ❏ Does the state GIS coordinator have around-the-clock contact information for his or her emergency management counterpart and his or her backup?
- ❏ Do the state GIS coordinator and the state emergency management counterpart (and their respective backups) hold regular meetings to determine any gaps in their geospatial support for your emergency management operations?
- ❏ Have action plans been developed to bridge those gaps?
- ❏ Have you established agreements with adjoining jurisdictions and with state and federal governments to share data and products?
- ❏ Have you established agreements with adjoining jurisdictions and with state and federal governments that determine what data and tools will be used during an emergency?
- ❏ Have you developed agreements between geospatial professional teams at the municipal, state, and federal levels that identify the roles that each level will play and who will produce what in order to avoid duplication of effort during a large event?
- ❏ (For the state emergency management agency) Have you worked with the state GIS coordinator to develop an inventory with around-the-clock contact information for GIS coordinators, their emergency management counterparts, and their respective backups in each county or major municipality in your state?
- ❏ (For the state emergency management agency) Has this information been distributed to the emergency management community and the GIS coordinators in each county or major municipality in your state?
- ❏ (For FEMA) Have you developed a secure web site with around-the-clock contact information for GIS coordinators, their emergency management counterparts, and their respective backups in each state?
- ❏ (For FEMA) Has this information been distributed to the federal agencies that respond to emergencies as well as to the emergency management community and the GIS coordinators in each state?

Human Resources

- ❏ Do you have a designated geospatial team that is regularly deployed in your EOC for emergencies?
- ❏ Have you developed an organizational structure for your team that defines the roles of team members (manager, liaison, and technical support staff)?
- ❏ Does your organization have a geospatial team (away team) that you can deploy to incident sites to assist in emergency response?
- ❏ Does your organization have a geospatial modeling team established, with scientific expertise in developing various models for plume analysis, hurricane surges, flooding, etc.?
- ❏ (For state GIS coordinators and the state emergency management agency) Have you worked together to develop a list of additional geospatial professionals (volunteers) in your state (along with their areas of expertise and around-the-clock contact information) who can be called upon to assist in an emergency?
- ❏ (For state GIS coordinators and the state emergency management agency) Have you worked together to develop a secure web site to distribute this information to authorized users?

Training

- ❏ Is the use of geospatial data and tools included as part of your emergency training exercises?
- ❏ Are these exercises conducted more than once a year?
- ❏ Do your emergency response professionals understand the capabilities that geospatial data and tools offer to improve their ability to plan for and respond to incidents?
- ❏ Have you established a training program for your first responders and emergency management decision makers on the use of geospatial data and tools in their workflow and decision-making processes?
- ❏ Are the first responders and emergency management decision makers trained on this at least once a year?
- ❏ Have you established a training program for your geospatial team in emergency management organization concepts and operational procedures?
- ❏ Are these responders trained on this at least once a year?
- ❏ Have you established a training program for your geospatial team in the use of geospatial data and tools during a disaster?
- ❏ Is the team trained on this more than once a year?

- Does your geospatial team train with predeveloped map templates?
- Do you conduct scenario-based training exercises that include geospatial professionals and the use of geospatial data and tools in the emergency management work cycle and decision-making process?
- Are the geospatial professional team manager and liaison included in the scenario training exercise meetings and briefings to allow them to understand better how geospatial data and tools are being used in the decision-making process?
- Do you conduct these exercises on a quarterly basis at a minimum?
- Are the results of these exercises posted to a secure web site so that other authorized responders not involved in the exercise can learn from them?
- Have you integrated the use of an on-site geospatial unit (away team) into your training program?
- Has your geospatial modeling team been incorporated into your scenario training exercises?

Data Access
- Have you developed relationships through the state GIS coordinator with the data custodians and established protocols and agreements, where required, to ensure access to and use of the data you require for planning, training, and emergency response activities?
- Have you developed a methodology to ensure regular updates to those data?
- Are your geospatial data backed up on a regular basis?
- Do you have a full copy of the data?
- Do you have copies of the data securely stored in different geographic regions of your state?
- Do you have a copy of the data securely stored in a different state or geographic region of the country?
- Have you tested your methodologies for rebuilding your servers using the backed-up data within the past year?
- Have you tested the process for accessing data from data-sharing partners during simulations to ensure the viability of your methodology?
- Have you established a web-based GIS service to encourage rapid access to and delivery of event-based data?
- (For FEMA) Have you worked with the state GIS coordinators to

APPENDIX C   167

  develop a secure web site within each state with an inventory
  (with around-the-clock contact information for the data custodians) of geospatial data in each state for use in emergency management operations?
- ❏ (For FEMA) Have you developed links to each of these state inventories and made this resource available to local, county, state,
  and federal agencies that would respond to a catastrophe?

Data Quality

- ❏ Do you have geospatial data on your critical infrastructure? Do
  they include the following:

  - ❏ Utilities (water, sewer, electric, gas, and petroleum lines and
    their related facilities);
  - ❏ Telecommunications lines including phones, networks, and
    cable;
  - ❏ Cell and other communication towers;
  - ❏ Transportation systems;
  - ❏ Shelters;
  - ❏ Dams;
  - ❏ Petroleum and chemical storage sites;
  - ❏ Hazardous waste sites;
  - ❏ Fire departments;
  - ❏ Police departments;
  - ❏ EMS (emergency medical service) districts;
  - ❏ Ambulance services;
  - ❏ Educational facilities;
  - ❏ Medical facilities;
  - ❏ Government facilities;
  - ❏ Military facilities;
  - ❏ Religious facilities;
  - ❏ Long-term care facilities;
  - ❏ Nursing homes;
  - ❏ Day care centers;
  - ❏ Animal control facilities; and
  - ❏ Animal shelters?

In addition, do you have:

  - ❏ Imagery;
  - ❏ ZIP code boundary data;
  - ❏ Roads and address data;

- Elevation data;
- Flood zones;
- Property data;
- Aquifers and other hydrological features;
- Data that locate businesses and industry in your state and detail their numbers of employees;
- Census data;
- Data on the daytime populations in specific areas;
- Data on the agricultural industry, including types of crops or animals housed at each site;
- Data on emergency equipment (pumps, generators, cots, blankets, etc.) or supplies (water, food, etc.) that are ready for deployment during an emergency; and
- Data from surrounding regions or states?

- Has your geospatial data team determined the quality and usability of the geospatial data gathered for emergency response?
- Do the metadata provide an adequate description of data quality, including accuracy and currency?

Data Gathering

- Have you established a team to identify and gather all geospatial data needed for your emergency response activities?
- Has your geospatial data team determined the quality and usability of the geospatial data gathered for emergency response?
- Have you worked with your state GIS coordinator to develop an inventory of municipal, county, state, and federal data that you require for use in emergency response?
- Does this inventory include around-the-clock contact information for the data custodian or owner?
- Does this inventory include metadata documenting and describing the geospatial data?
- Does your state have contracts in place for emergency aerial imagery?
- Do you have agreements in place to acquire digital images via government or private-sector helicopter, etc., of event sites immediately after an event occurs?
- Do you have agreements in place and near-live data feeds from utilities detailing the geographic extent of power outages?
- Do you have live or near-live geospatial weather data?
- Do you have live or near-live geospatial data on road conditions and capacities or other transportation systems?

- ❏ Do you have any near-live data feeds from hospitals or other medical facilities detailing geospatial data on bed capacity or medication availability?
- ❏ Do you have the capability to track the distribution of your emergency equipment or supplies geographically?
- ❏ Have you tested your data-gathering methodologies in training exercises?
- ❏ Do you have a geospatial web-based service application that provides rapid access to your event-related data by regional, state, or federal organizations responding to a large event?

Data Improvement

- ❏ Has the geospatial data team identified which data require improvements and which data not currently available need development?
- ❏ Has this team worked with the state GIS coordinator to coordinate the required work?
- ❏ Do you get updates to your data (not including imagery) on an annual basis at a minimum?
- ❏ Is the imagery for your state less than five years old?
- ❏ Do you have a system for improving geospatial data to meet your emergency response requirements?
- ❏ (For FEMA) Have you worked through the state GIS coordinators to establish a program to identify the needs in each state for data improvement and the creation of new data where none exist?
- ❏ (For FEMA) Have you developed a mechanism coordinated with the state GIS coordinator to provide funding and resources to assist state and local governments in improving and developing those data?

Information Delivery

- ❏ Has your geospatial team practiced rapid delivery of geospatial information to meet emergency management decision-making requirements? Can it deliver standard products required by your emergency managers within two hours of an event?
- ❏ Has your geospatial team developed templates to improve the speed of delivery of geospatial information during an emergency?
- ❏ Have you developed models depicting the impact of hurricanes or floods on the community, based on the category of the hurricane and the projected height of the flood waters?

- Do you have an easy-to-use on-line application that allows emergency responders at all levels who are not geospatial professionals to make geospatial inquiries to resolve issues?
- Do you have automated geocoding capabilities that will allow your geospatial team (or nontechnical staff) to convert address locations to latitude and longitude quickly, to assist rescuers during disasters (such as floods) in locating individuals in need of rescue?
- Are your requests for assistance during an emergency tracked in a database and tracked via a GIS application to provide visual analysis of problem patterns, etc.?
- Have your geospatial professionals developed agreements with geospatial professional teams in adjacent communities or the state, and at the federal level, to determine the data and tools to be used and shared during disasters?
- Have your geospatial professionals developed agreements with geospatial professional teams in adjacent communities or the state, and at the federal level, on the roles that each level will play and the products that will be generated in order to avoid duplication of effort during a disaster?

Equipment and Infrastructure

- Do you have up-to-date geospatial software and hardware in your EOC?
- Do you have electronic field data collection methods (using Global Positioning System [GPS] enabled handheld computers with wireless communication systems) available to determine the geographic extent of an incident?
- Do you have capabilities of obtaining digital photographs of incident sites and transmitting them wirelessly to the EOC?
- Does your state have geospatial equipment and data prepared for deployment near an incident site?
- Do you have a vehicle (van, recreational vehicle, etc.) that has hardware, GIS software, data, and wireless communication systems installed and prepared for deployment to an incident site?
- Has your staff trained in this vehicle during scenario-based exercises?
- Do you have the ability to push out or pull in geospatial data or web-based services across the Internet?
- Do you have backup satellite communications systems to transmit geospatial data when necessary?

- ❏ (For the state emergency management agency and the state GIS coordinator) Have you worked together to develop an up-to-date inventory of geospatial hardware available for use in an emergency (and around-the-clock contact information) in your state?
- ❏ (For FEMA) Have you developed a secure web site with this inventory and around-the-clock contact information for each state?
- ❏ (For FEMA) Have you developed a secure, national GIS web-based application to enable data to be accessed by authorized users across the country?

Appendix D

Workshop Agenda and Participants

**WORKSHOP ON GEOSPATIAL INFORMATION
FOR DISASTER MANAGEMENT**

Agenda

Wednesday, October 5, 2006

8:30 a.m.   Continental Breakfast

9:00   Welcome (Goodchild)

PANEL 1: User Needs: Requirements and Gaps
9:10   Introduction (Donahue, Bruzewicz)
9:15   Panelist Remarks
             Questions by Committee
             Open Discussion
10:35   Summary of Discussion (Donahue, Bruzewicz)

10:40   **Break**

PANEL 2: Data and Tools: Requirements and Gaps
11:00   Introduction (Goodchild, Cutter)
11:05   Panelist Remarks
             Questions by Committee
             Open Discussion

12:25 p.m. Summary of Discussion (Goodchild, Cutter)

12:30-1:30 **Lunch**

PANEL 3: Interoperability
1:30   Introduction (Moeller, Hu, Reed, Klavans)
1:35   Panelist Remarks
            Questions by Committee
            Open Discussion
2:55   Summary of Discussion (Moeller, Hu, Reed, Klavans)

3:00   **Break**

BREAKOUT SESSION
3:30-5:00
1—Members Room
2—Room 142
3—Room 146

PLENARY SESSION
5:00   Summary of Breakout Sessions
5:30   Adjourn

5:30   **Reception in Great Hall**

Thursday, October 6, 2006

Venue: Members Room

8:30 a.m.   Continental Breakfast

PANEL 4: Training
9:00   Introduction (Stanley, Cutter)
9:05   Panelist Remarks
            Questions by Committee
            Open Discussion
10:25   Summary of Discussion (Stanley, Cutter)

10:30   **Break**

APPENDIX D

PANEL 5: Data Accessibility and Security
11:00   Introduction (Gomez, Monmonier, Oswald)
11:05   Panelist Remarks
        Questions by Committee
        Open Discussion
12:25 p.m. Summary of Discussion (Gomez, Monmonier, Oswald)

12:30   **Lunch**

BREAKOUT SESSION
1:30-3:00
1—Members Room
2—Room 142
3—Room 146

3:00   **Break**

PLENARY SESSION
3:30   Reports from Breakout Session
4:00   Wrap-up of Workshop
4:30   Adjourn

Session Panel Members and Chairs

Session 1: User Needs: Requirements and Gaps

Amy Donahue, University of Connecticut (co-chair)
Andy Bruzewicz, U.S. Army Corps of Engineers (co-chair)
Jim McConnell, New York City Office of Emergency Management
Michael Payne, Pierce County, Washington
Suha Ulgen, United Nations Office for the Coordination of
   Humanitarian Affairs
Bruce Davis, Department of Homeland Security

Session 2: Data and Tools: Requirements and Gaps

Mike Goodchild, University of California, Santa Barbara (co-chair)
Susan Cutter, University of South Carolina (co-chair)
Paul Densham, University College London (co-chair)
Michael Hodgson, University of South Carolina
Tim Johnson, North Carolina Center for Geographic Information and
   Analysis
Bob Chen, Columbia University

Stephen Smith, Oak Ridge National Laboratory
Earnie Paylor, WorldTech Inc.
John Perry, Federal Emergency Management Agency (FEMA)
Charles Huyck, ImageCat

Session 3: Interoperability

John Moeller, Northrup Grumman TASC (co-chair)
Pat Hu, Oak Ridge National Laboratory (co-chair)
Judith Klavans, University of Maryland (co-chair)
Carl Reed, Open Geospatial Consortium (co-chair)
Judith Woodhall, COMCARE
Craig Stewart, GeoConnections Secretariat, Natural Resources Canada
Tom Merkle, Capital Wireless Integrated Network
Tony Spicci, Missouri Department of Conservation
John Contestabile, Maryland Department of Transportation
Eric Anderson, City of Tacoma, Washington

Session 4: Training

Ellis Stanley, Los Angeles Emergency Preparedness Department (co-chair)
Susan Cutter, University of South Carolina (co-chair)
Sue Kalweit, Booz Allen Hamilton
David McEntire, University of North Texas
Robert Slusar, Northrup Grumman IT TASC
Ron Wilson, National Institute of Justice
John Hwang, California State University, Long Beach

Session 5: Data Accessibility and Security

Pete Gomez, Xcel Energy (co-chair)
Mark Monmonier, Syracuse University (co-chair)
Bruce Oswald, New York State (retired) (co-chair)
Don Welsh, New York State Office of Cyber Security and Critical Infrastructure Coordination
Peter Gomez, Xcel Energy
Beth Lachman, Rand Corporation
Harlan Onsrud, University of Maine

Appendix E

Biographical Sketches of Committee Members and Staff

Michael F. Goodchild (NAS) is a professor of geography at the University of California, Santa Barbara, and chair of the Executive Committee of the National Center for Geographic Information and Analysis. He received a B.A. in physics from Cambridge University and a Ph.D. in geography from McMaster University. He taught at the University of Western Ontario for 19 years before moving to his present position in 1988. His research interests focus on the generic issues of geographic information, including accuracy and the modeling of uncertainty, the design of spatial decision support systems, the development of methods of spatial analysis, and data structures for global geographic information systems (GIS). He has received several awards and published numerous books and journal articles. He is a member of the National Academy of Sciences, and a member of the National Research Council's (NRC's) Geographical Sciences Committee, and was a member and a chair of the NRC's Mapping Science Committee (1992-1999).

Andrew J. Bruzewicz is director of the U.S. Army Corps of Engineers (USACE) Remote Sensing/GIS Center and has served as the program manager for the Corps' civil works geospatial research and development program since its inception in 1998. Since 1991, he has conducted research on the application of geospatial technologies to disaster preparedness, planning, and response, and since 1999, he has been integrating these technologies into the USACE emergency management processes. Mr. Bruzewicz is presently the team leader for the Corps' GIS Planning and

Response Team and serves as the Corps' liaison to the Federal Emergency Management Agency (FEMA) for geospatial data collection and sharing during disasters. He is the past national education chair of the American Society for Photogrammetry and Remote Sensing and has presented at two National Academies workshops. Mr. Bruzewicz holds an A.B. in economics and an A.M. in geography from the University of Chicago.

Susan L. Cutter is the director of the Hazards Research Lab, a research and training center that integrates geographical information processing techniques with hazards analysis and management, as well as a Carolina Distinguished Professor of Geography at the University of South Carolina. She received her Ph.D. in geography from the University of Chicago. She is the co-founding editor of an interdisciplinary journal *Environmental Hazards*, published by Elsevier. She has worked in the risk and hazards fields for more than 25 years and is a nationally recognized scholar in this field. She has authored or edited eight books and more than 50 peer-reviewed articles. Dr. Cutter is a fellow of the American Association for the Advancement of Science (AAAS), was president of the Association of American Geographers (1999-2000), and is a member of the NRC's Geographical Sciences Committee.

Paul J. Densham is a reader in geography and a researcher in the Centre for Advanced Spatial Analysis at University College London (UCL). His research interests and publications focus on spatial decision support systems, locational analysis, GIS, and parallel algorithms for spatial problems. Prior to joining UCL, he was an assistant professor of geography at the State University of New York at Buffalo and a research fellow in the U.S. National Center for Geographic Information and Analysis (NCGIA). He co-led NCGIA's research initiatives Spatial Decision Support Systems and Collaborative Spatial Decision Making, and led the investigation Parallel Computation and GIS. Dr. Densham has applied spatial decision support systems in his work on dynamic location strategies for emergency service vehicles (with Cadcorp, Ltd.) and other work with the Environmental Systems Research Institute, HSBC, NYNEX Science and Technology, Iowa Department of Education, and planning offices and government shops in India; he has also worked on migration and biodiversity problems. He holds a B.A. in geography and economics from the University of Keele, an M.Sc. in operational research from the University of Birmingham, and a Ph.D. in geography from the University of Iowa.

Amy K. Donahue is associate professor of public policy at the University of Connecticut. Dr. Donahue's research focuses on the productivity of emergency response organizations and on the nature of citizen demand for public safety. For the past two years, Dr. Donahue has served as a technical adviser to the Department of Homeland Security's (DHS's) Science and Technology Directorate, helping to develop research and development programs to meet the technological needs of emergency responders. From 2002 to 2003, Dr. Donahue served as senior adviser to the Administrator for Homeland Security at the National Aeronautics and Space Administration (NASA). She was the agency's liaison with DHS and the Homeland Security Council and identified opportunities for NASA to contribute to homeland security efforts across government. In 2003, Dr. Donahue spent three months in Texas helping manage the *Columbia* recovery operation, an intergovernmental response that involved 450 organizations and 25,000 responders. Prior to her affiliation with the University of Connecticut, Dr. Donahue was a senior research associate at the Alan K. Campbell Public Affairs Institute at Syracuse University. She also has 20 years of training and field experience in an array of emergency services-related fields, including managing a 911 communications center and working as a firefighter and emergency medical technician in Fairbanks, Alaska, and upstate New York. As a distinguished military graduate of Princeton's Reserve Officer Training Corps, she served on active duty for four years in the 6th Infantry Division, rising to the level of captain. Dr. Donahue holds her Ph.D. in public administration and her M.P.A. from the Maxwell School of Citizenship and Public Affairs at Syracuse University. Her B.A. from Princeton University is in geological and geophysical sciences.

J. Peter Gomez is manager of information requirements for Xcel Energy. He has broad knowledge and experience of critical infrastructure protection, particularly in the context of gas and electric system planning and design, asset management, and data system integration. Mr. Gomez has been with Xcel Energy (formerly the Public Service Company of Colorado) since 1985. His current responsibilities include the expansion of the corporation's asset management system across the company's 10-state service territory. He also serves as a primary business liaison between Xcel Energy's operations organization and IBM, the information technology business partner of Xcel Energy. Earlier, he led Xcel Energy's Geographic Information System and Outage Management System application development organization. Mr. Gomez received a B.S. degree in mechanical engineering from New Mexico State University. He has served on advisory boards for the Colorado School of Mines, Metro State University, and the Denver Public Schools Career Development Center. Since 2000, Mr. Gomez has been on the Board of Directors of the Geospatial Information Technology Association (GITA) and served as its president in 2005.

Patricia Hu is director of the Center for Transportation Analysis at the Engineering Science and Technology Division of Oak Ridge National Laboratory (ORNL). She has been at ORNL since 1982 and in her current position since 2000. Ms. Hu contributed to DHS' research and development plan for critical infrastructure protection. Additionally, she led a team (supported by the Transportation Security Administration) that studied the domain awareness of U.S. food supply chains by linking and analyzing geospatial data on transportation networks, traffic volume, choke points and congestion, freight flow, and traffic routing. Ms. Hu holds a B.C. from the National Chengchi University, Taipei, Taiwan, and an M.S. in mathematics and statistics from the University of Guelph, Guelph, Ontario, Canada. Currently, she chairs the NRC's Transportation Research Board (TRB) Committee on National Data Requirements and Programs and has been a member of more than 10 other TRB committees.

Judith Klavans is director of research at the Center for Advanced Study of Language in the College of Information Studies, University of Maryland. Her research is in digital libraries, linguistics, and natural language systems. Until 2004, she was director of the Center for Research on Information Access at Columbia, which is responsible for linking theoretical computer science research with operational applications such as digital libraries and digital government. She is a principal investigator in several large projects, including the National Science Foundation (NSF) funded PERSIVAL medical digital library, the NSF and Bureau of Labor Statistics (BLS) supported Digital Government Research Center joint with University of Southern California-ISI, the TIDES (Translingual Information Detection, Extraction, and Summarization) multilingual summarization project funded by the Defense Advanced Research Projects Agency, and most recently, the Mellon-supported CLiMB (Computational Linguistics for Metadata Building) project, which links text and image collections. Prior to arriving at Columbia, Dr. Klavans spent nearly 10 years in the Computer Science Division of the T.J. Watson IBM Research Division, where her work included extracting information from machine-readable dictionaries, building bilingual aligned phrasal dictionaries, and text-to-speech. Earlier, she was postdoctoral fellow at the Massachusetts Institute of Technology in linguistics and computer science. Dr. Klavans holds a Ph.D. and an M.A. in linguistics from the University of London, an M.Ed. in English as a second language from Boston University, and a B.A. in Spanish and mathematics from Oberlin College.

John J. Moeller is senior principal engineer at Northrop Grumman TASC (The Analytical Sciences Corporation). He is senior adviser for geospatial interoperability, geospatial architectures and infrastructures, and national and international geospatial policy and strategic issues. Prior to joining TASC in 2002, Mr. Moeller accumulated 34 years of federal government management and leadership experience, including 7 years as staff director of the Federal Geographic Data Committee, 24 years in leadership roles at the Bureau of Land Management, and 3 years as an officer in the U.S. Army. Mr. Moeller has received numerous awards and letters of commendation for outstanding performance, and received the Federal Computer Week 2002 Federal 100 award presented to 100 leaders who made a difference in federal information technology during 2001. He has a B.S. in forestry from the University of New Hampshire and an M.S. in natural resource management from the State University of New York.

Mark Monmonier is the distinguished professor of geography at Syracuse University. Dr. Monmonier has used geographic information systems extensively as a tool in estimation and communication of risk and uncertainty with natural and technological hazards. In addition, he researches the use of multimedia and other advanced technologies in the design and use of maps throughout society as analytical and persuasive tools in homeland security, journalism, politics, public administration, and science. Furthermore he researches the legal and ethical issues in intellectual property, liability, privacy, and public access and in mapping policy at the state and national levels. He is the author of nine books, including *How to Lie with Maps*. Dr. Monmonier has received numerous awards, including the Globe Book Award for Public Understanding of Geography, and awards from the Canadian Cartographic Association, the American Geographical Society, the Pennsylvania Geographical Society, and the Association of American Geographers. Dr. Monmonier was a member of the NRC's Mapping Science Committee from 1999 to 2005. He has a B.A. from the Johns Hopkins University and an M.S. and Ph.D. from the Pennsylvania State University.

Bruce Oswald is vice president of public sector geospatial solutions for the James W. Sewall Company. Mr. Oswald leads the company's Public Sector GIS Consulting Group, where he directs Sewall projects and initiatives at the statewide and state-agency level. Founded in 1880, Sewall has provided comprehensive GIS consulting services to government, utility companies, and the forest industry since the 1970s. Prior to joining Sewall, Mr. Oswald was the assistant director and chief information officer of the New York State (NYS) Office of Cyber Security and Critical Infrastructure Coordination. There he was responsible for the implementation of New York's cyber security and statewide GIS programs, as well as its critical infrastructure coordination efforts in response to the terrorist attacks on the World Trade Center. Mr. Oswald also served as the chair of the NYS

GIS Coordinating Body. He has been involved in the use of GIS for emergency response for major snow and ice storms, hurricanes, landslides, and potential and real terrorist events since 1998 and in the development of New York's new Critical Infrastructure Application, which provides access to geographic information technology for nontechnical emergency responders. Prior to serving in that position, Mr. Oswald served as the assistant director for statewide initiatives within the NYS Office for Technology as well as the director of the Center for Geographic Information. Mr. Oswald is a certified project management professional and a licensed landscape architect. He holds B.S. degrees in environmental science and landscape architecture from the College of Environmental Science and Forestry at Syracuse University, and an M.B.A. from Rensselaer Polytechnic Institute.

Carl Reed is currently the chief technology officer of the Open Geospatial Consortium, Inc. (OGC), a nonprofit trade association with a current membership of 270 commercial, government, and academic organizations whose primary objective is to create a consensus forum and related industry collaboratives to solve technical and business development problems relating to geoprocessing. Prior to the OGC, Dr. Reed was the vice president of geospatial marketing at Intergraph. This was after a long tenure at the GIS software company Genasys II, where he served as chief technology officer for Genasys II worldwide. From 1989 to 1996, Dr. Reed was president of the Genasys U.S. operation. Before his tenure at Genasys, Dr. Reed worked at Autometric for six years as GIS division manager, developing a variety of systems for the civilian branches of the U.S. federal government as well as for the military. Dr. Reed has designed and implemented two major GIS packages, MOSS and GenaMap. Dr. Reed received his Ph.D. in geography, specializing in GIS technology, from the State University of New York at Buffalo in 1978. In 1996, in recognition of his contributions to the GIS industry, he was voted one of the top 10 most influential people in the industry.

Ellis M. Stanley, Sr., is the general manager of the City of Los Angeles Emergency Preparedness Department. He has directed emergency management programs around the United States for 25 years and has also served as a county fire marshal, fire and rescue commissioner, and county safety officer. He has a B.S. in political science from the University of North Carolina at Chapel Hill. Mr. Stanley was president of the International Association of Emergency Managers, the American Society of Professional Emergency Planners, and the National Defense Transportation Association. He chaired the Certified Emergency Managers Certification Commission, is vice-chair of the Association of Contingency Planners, and is vice-president for the public sector of the Business & Industry Council on Emergency Preparedness and Planning. Additionally, Mr. Stanley is on the Emergency Services Committee of the American Red Cross Los Angeles Chapter, the Emergency Preparedness Commission for the County and Cities of Los Angeles (currently as chairman of the board), the City's Emergency Operations Board, and the board of directors of the National Institute of Urban Search and Rescue. He has many other emergency management-related advisory and training roles. As a trainer, Mr. Stanley served as the discipline expert for major sporting events (Olympics, Special Olympics, World Cup, Super Bowls, World Series, National Basketball Association Championships), national political conventions, and the Papal and World Youth Conference in Denver in 1994. He is the City of Los Angeles' representative to the Cluster Cities Project of the Earthquake Mega-cities Initiative—a project that fosters sharing of knowledge, experience, expertise, and technology to reduce risk to large metropolises from earthquakes and other major disasters. Mr. Stanley is also an adviser to the Multidisciplinary Center for Earthquake Engineering Research. He is a member of the NRC's Natural Disasters Roundtable.

National Research Council Staff

Ann G. Frazier is a program officer with the Board on Earth Sciences and Resources, coordinating mapping science activities. Prior to working for the NRC, Ann had 23 years of experience in science and engineering, including 10 years with the U.S. Geological Survey (USGS) in geographic sciences. She focused on land-cover change, urban growth, ecological modeling, and application of geographic analysis and remote sensing in interdisciplinary environmental studies. Prior to the USGS, Ann worked for 13 years in the aerospace industry on the Space Shuttle and Space Station Programs. Ann has a B.A. in physics-astronomy, an M.S. in space technology, a certificate in environmental management, and an M.S. in geography.

Amanda M. Roberts is a senior program assistant with the Board on Earth Sciences and Resources. Before coming to the National Academies, she interned at the Fund for Peace in Washington D.C., working on the Human Rights and Business Roundtable. Amanda also worked in Equatorial Guinea, Africa, with the Bioko Biodiversity Protection Program. She is a master's student at the Johns Hopkins University in the environment and policy program and holds an M.A in international peace and conflict resolution from Arcadia University, specializing in environmental conflict in sub-Saharan Africa.

Jared P. Eno is a senior program assistant with the Board on Earth Science and Resources. Before coming to the National Academies, he interned at Human Rights Watch's Arms Division, working on the 2004 edition of the *Landmine Monitor Report*. Jared received his A.B. in physics from Brown University.